The City & Guilds textbook

Painting and Decorating

LEVEL 1 DIPLOMA (6707)
LEVEL 2 TECHNICAL CERTIFICATE (7907)
LEVEL 2 DIPLOMA (6707)

Barrie Yarde
Steve Olsen

HODDER
EDUCATION
AN HACHETTE UK COMPANY

Orders: please contact Bookpoint Ltd, 130 Park Drive, Milton Park, Abingdon, Oxon OX14 4SE. Telephone: +44 (0)1235 827827. Fax: +44 (0)1235 400401. Email education@bookpoint.co.uk Lines are open from 9 a.m. to 5 p.m., Monday to Saturday, with a 24-hour message answering service. You can also order through our website: www.hoddereducation.co.uk

ISBN: 978 1 3983 0577 9

© The City and Guilds of London Institute and Hodder & Stoughton Limited 2020

First published in 2020 by
Hodder Education,
An Hachette UK Company
Carmelite House
50 Victoria Embankment
London EC4Y 0DZ

www.hoddereducation.co.uk

Impression number   10 9 8 7 6 5 4 3 2 1

Year   2024 2023 2022 2021 2020

Cover photo © Tomasz Zajda – stock.adobe.com

City & Guilds and the City & Guilds logo are trade marks of The City and Guilds of London Institute. City & Guilds Logo © City & Guilds 2020

Typeset by Integra Software Serv. Ltd.

Printed in Slovenia.

A catalogue record for this title is available from the British Library.

Contents

About the authors

Barrie Yarde

I have been very fortunate to have had a long career in construction, and in particular as a painter and decorator. I was trained as an apprentice by an 'old master', who helped me gain the skills that have given me such a wonderful career.

It was very much a case of practice and more practice, until I was able to demonstrate mastery of the skills that I had been shown.

I later went into teaching, and the years spent teaching in college have, I believe, delivered the right result – many of my learners have developed into fine craftspeople themselves.

I hope I can continue to encourage others to follow a worthwhile career in construction. I have been inspired by many others – too numerous to mention – and so I hope that this textbook will provide inspiration for you, encouraging you to study and learn the skills of painting and decorating.

Steve Olsen

I am a strong supporter of apprenticeships and training programmes for young and old learners alike. I left school in 1984 to serve an apprenticeship in Painting and Decorating for Swan Hunter Shipbuilders, Wallsend, Newcastle and have always been employed as a result of the qualifications I gained during that time and being classed as a tradesman.

Being a Painter and Decorator has given me more opportunities than I could have hoped for. At first, I progressed to being a charge-hand within the industry and then I became an assessor and a lecturer delivering technical and NVQ qualifications to learners within a private training provider. I now work as a Construction Lecturer and Curriculum Lead at an FE college. I have also worked as an Internal Quality Assurer and External Quality Assurer carrying out compliance work within the education sector, written training books and been involved in qualification development work with City & Guilds. None of this would have been possible without my qualifications as a Painter and Decorator.

Work hard and be committed to your learning/training because there will be more opportunities available for you in the future!

I would like to take this opportunity to thank my wife, Lynne, my daughters, Christy and Lydia, and my mother, father, brothers and their families, as well as my close friends for their love and support that I received during a difficult time in my life. I was diagnosed with Stage 4 Throat Cancer in August 2016 and would not have survived this without their help.

How to use this book

Throughout this book you will see the following features:

Qualification Mapping

At the start of each chapter, there is a qualification mapping grid which indicates how the content relates to each of the learning outcomes. The layout of the book has been mapped to the CG 7907 L2 Technical Diploma covers the main points of all qualifications.

INDUSTRY TIPS are particularly useful pieces of advice that can assist you in your workplace.

> ## INDUSTRY TIP
>
> Stretch string or a rubber band across the middle of the top of the bucket when pasting. This gives you somewhere to rest the brush when not in use.

KEY TERMS in **bold purple** in the text are explained to aid your understanding. (They are also explained in the Glossary at the back of the book.)

> ## KEY TERM
>
> **Defect-free:** without flaws, holes or cracks and bits left on the surface.

HEALTH AND SAFETY boxes flag important points to keep yourself, colleagues and clients safe in the workplace. They also link to sections in Chapter 1 for you to recap learning.

> ## HEALTH AND SAFETY
>
> Some paints give off fumes as they dry, so make sure that your work area is well ventilated.

ACTIVITIES help to test your understanding and learn from your colleagues' experiences.

> ## ACTIVITY
>
> Explain why a stair scaffold should be used for staircases in preference to a ladder and plank. Refer to the HSE website regarding the Work at Height Regulations 2005 (as amended).

IMPROVE YOUR MATHS items combine improving your understanding of painting and decorating with practising or improving your maths skills.

IMPROVE YOUR ENGLISH items combine improving your understanding of painting and decorating with practising or improving your English skills.

At the end of each chapter there are some TEST YOUR KNOWLEDGE questions and PRACTICAL TASKS. These are designed to identify any areas where you might need further training or revision. Answers to the questions are at the back of the book.

Acknowledgements

This book draws on several earlier books that were published by City & Guilds, and we acknowledge and thank the writers and editors of those books:

- Ann Cook
- Colin Fearn
- Steve Walter
- Martin Burdfield.

We would also like to thank everyone who has contributed to City & Guilds photoshoots. In particular, thanks to: Jules Selmes and Adam Giles, Steve Lammas, Gary Thoirs, Casie Bedwell, Hamed Bamba, Lee Farrell, Luke Kalavashoti, Jade Keen, Stefan Kuhl, Andrew Maughan, Ismail Mohamud, Kris O'Neill, Abdul Qaffar, Mohammed Sanusi-Omosanya, Alan Saxton, James Shiels and Billy Snowball.

Text credits

Photo credits

© Clow Group Ltd; r © Natara/stock.adobe.com; **p.71** l © 300dpi/stock.adobe.com; r © Simon Turner/Alamy Stock Photo; **p.72** l © Clow Group Ltd; r © Semen Sullivanchuk/stock.adobe.com; **p.73** tl © Youngman staging board/Werner UK Sales & Distribution Ltd; bl © Chase Manufacturing Ltd; **p.74** tl © Annie Eagle/Alamy Stock Photo; r © haraldmuc/Shutterstock; **p.75** l © Werner UK Sales & Distribution Ltd. Image Name: BoSS Tower User Guide cover; r © City & Guilds; **p.77** br © Pat Labels Online; **p.80** For tower safety advice, visit pasma.co.uk; **p.81** tl © UrbanImages/Alamy Stock Photo; **p.84** tr © Australian Scaffold; br © bikeriderlondon/Shutterstock; **p.85** tl © City & Guilds; tr © City & Guilds; br © ChiccoDodiFC/stock.adobe.com; **p.86** l © 7505811966/Shutterstock; r © Panitan/stock.adobe.com; **p.94** t © Semjonow Juri/Shutterstock; m © Barnaby Chambers/Shutterstock; b © arigato/Shutterstock; **p.95** t © Karakedi35/stock.adobe.com; m © Peter Guess/Shutterstock; b © Dmytro Synelnychenko/stock.adobe.com; **p.96** t © jocic/Shutterstock; m © joeystudio/Shutterstock; **p.97** t © vaaka/Shutterstock; b © Waldenstroem/stock.adobe.com; **p.98** t © Patryk Michalski/Shutterstock; m © Alexey V Smirnov/Shutterstock; b © Alexus/stock.adobe.com; **p.99** t © Andrey_Maksimov/stock.adobe.com; tm © Sascha Preussner/Shutterstock; bm © Patita88/stock.adobe.com; b © Alexanian Arsen/stock.adobe.com; **p.100** t © Ratikova/Shutterstock; m © Koxae/stock.adobe.com; b © Richard Nixon Cover Images/Alamy Stock Photo; **p.102** © Unkas Photo/stock.adobe.com; **p.103** © City & Guilds; **p.104** l © juriskraulis/stock.adobe.com; r © MelissaMN/stock.adobe.com; **p.105** © City & Guilds; **p.107** © Avalon/Construction Photography/Alamy Stock Photo; **p.108** © Everbuild - A Sika Company; **p.109** t © Candus Camera/Shutterstock; rest © City & Guilds; **p.111** t © Mirror-images/stock.adobe.com; m © Heather M Greig/Shutterstock; b Photo by Huntco; **p.112** t © Christopher Elwell/Shutterstock; tm © eltoro69/Shutterstock; bm © Ivan Kebe photography/Shutterstock; b © David Reilly/Shutterstock; **p.113** t © Kunal Mehta/Shutterstock; tm © debbie winchester/Shuttesrstock; bm https://commons.wikimedia.org/wiki/File:Eisenhammerschlag_Fe3_O4.jpg#/media/File:Eisenhammerschlag_Fe3_O4.jpg; http://www.gnu.org/licenses/old-licenses/fdl-1.2.html; b © Photomarine/Shutterstock.com; **p.114** t © Ilya Akinshin/Shutterstock; m © Alex Ishchenko/stock.adobe.com; b © Олег/stock.adobe.com; **p.115** t ©

TheFarAwayKingdom/stock.adobe.com; tm © Winiki/Shutterstock; m © Nigel Prosser/Alamy Stock Photo; bm © Faithfull Tools; b © Pawel/stock.adobe.com; **p.116** t Buckleys (UVRAL) Ltd – www.buckleysinternational.com; tm © Philip kinsey/stock.adobe.com; bm © mexrix/Shutterstock; b © City & Guilds; **p.117** l © Silo Stock/Alamy Stock Photo; r © Martyn F. Chillmaid/Alamy Stock Photo; **p.119** tl © Tuzemka/Shutterstock; bl © hadkhanong/Shutterstock; r © City & Guilds; **p.120** l © stocksolutions/Shutterstock; r © Ed Phillips/Shutterstock; **p.121** l © Nadiia/stock.adobe.com; r © Acclaimed Building Consultancy; **p.122** tl © City & Guilds; bl © The Concrete Countertop Institute; r © ChiccoDodiFC/Shutterstock; **p.123** l © Polina Katritch/Shutterstock; r © Ranee Sornprasitt/Alamy Stock Photo; **p.124** l © City & Guilds; r © imagedb.com/Shutterstock; **p.125** © Jeep5d/stock.adobe.com; **p.126** l © City & Guilds; **p.128** l © Valentin Agapov/Shutterstock; r © Tuzemka/Shutterstock; **p.129** t © izarizhar/Shutterstock; tm © hxdbzxy/Shutterstock; bm © bluehand/Shutterstock; b © City & Guilds; **p.130** © City & Guilds; **p.131** © Robertkoczera/stock.adobe.com; **p.132** © City & Guilds; **p.133** l © Anna/stock.adobe.com; **p.134** © City & Guilds; **p.135** © Shaun Daley/Alamy Stock Photo; **p.137** © City & Guilds; **p.138** l © Maryadam/stock.adobe.com; tr © Stanley Black and Decker; br © Dorling Kindersley ltd/Alamy Stock Photo; **p.140** © City & Guilds; **p.144** © BuildPix/Avalon/Construction Photography/Alamy Stock Photo; **p.145** tl © XAOC/Shutterstock; ml © Mindy w.m. Chung/Shutterstock; bl © Coprid/stock.adobe.com; tr © David Tran/stock.adobe.com; mr © Brilt/stock.adobe.com; **p.146** tl © DigitalGenetics/stock.adobe.com; bl © Faithfull Tools; r © Moodboard/stock.adobe.com; **p.147** l © endeavor/Shutterstock; r © City & Guilds; **p.148** l © Anton/stock.adobe.com; tr © City & Guilds; br FrogTape®; **p.149** © Shutter B Photo/Shutterstock; **p.150** l © City & Guilds; r © Robot recorder/stock.adobe.com; **p.151** l © huyangshu/Shutterstock; r © ANGHI/stock.adobe.com; **p.152** © Baloncici/Shutterstock; **p.154** l © Cranfield Colours Ltd; r © yuanann/Shutterstock; **p.156** l © Park Dale/Alamy Stock Photo; tr © Thaweesak/stock.adobe.com; br © Ekaterina43/Shutterstock.com; **p.162** t © Faithfull Tools; m © City & Guilds; b © Faithfull Tools; **p.163** t © City & Guilds; tm © City & Guilds; bm © Faithfull Tools; b © Iuliia Alekseeva/stock.adobe.com; **p.164** t © Coral Tools Ltd; m © City & Guilds; b © Faithfull Tools; **p.165** t © Faithfull Tools; ml © Michele

CHAPTER 1

PRINCIPLES OF CONSTRUCTION

INTRODUCTION

Construction is a vital part of the economy and plays an important role in all our lives. Working in this sector can be very rewarding and there are many opportunities for career progression. This chapter introduces the construction industry and gives a wider context to the decorating trade. This chapter will give you an understanding of the principles of construction, building technology and terminology. This includes legislation, such as health and safety, planning and building control.

By the end of this chapter, you will have an understanding of:
- how to work in the construction industry
- construction information
- how to set up and secure work areas
- building substructure
- building superstructure.

The table below shows how the main headings in this chapter cover the learning outcomes for each qualification specification.

Chapter section	Level 1 Diploma in Painting and Decorating (6707-13) Unit 101, 201	Level 2 Diploma in Painting and Decorating (6707-22/23) Unit 201, 202	Level 2 Technical Certificate in Painting and Decorating (7907-20) Unit 201	Level 3 Advanced Technical Diploma in Painting and Decorating (7907-30) Unit 301	Level 2 City & Guilds NVQ Diploma in Decorative Finishing and Industrial Painting Occupations (6572-20) Unit 101, 218, 608
1. Understand how to work in the construction industry	**101**: 7.1–7.5	**202**: 16.1–16.7	1.1–1.3	1.1–1.3	
2. Understand construction information	**201**: 1.1–1.5, 2.1–2.6, 3.1–3.5, 4.1–4.3, 5.1–5.3, 6.1–6.4, 7.1–7.4, 8.1–8.5, 9.1–9.4	**201**: 1.1–1.5, 2.1–2.6, 3.1–3.5, 4.1–4.3, 5.1–5.3, 6.1–6.4, 7.1–7.4, 8.1–8.5, 9.1–9.4 **202**: 10.1–10.2, 13.1–13.2	2.1–2.4	1.1–1.2, 4.1–4.2, 5.1–5.2, 6.1–6.3	**101**: 1.1–1.8, 2.1–2.6, 3.1–3.8, 4.1–4.3, 5.1–5.2 **218**: 1.1–1.3, 2.1–2.4, 3.1–3.3, 4.1–4.6 **608**: 1.1–1.4, 2.1–2.5, 3.1–3.6, 6.1–6.2, 7.1
3. Understand how to set up and secure work areas			3.1–3.3		
4. Know building substructure	**101**: 3.1–3.2, 5.1–5.3	**202**: 12.1, 14.1	4.1–4.2		
5. Know building superstructure	**101**: 4.1–4.3, 4.6, 6.1–6.3	**202**: 15.1	5.1–5.5		

1 UNDERSTAND HOW TO WORK IN THE CONSTRUCTION INDUSTRY

This section provides an overview of the range of construction activities. It includes the roles of members of a building team, communication methods used to share information and career progression within the industry.

▲ Figure 1.1 Members of the building team

Types of construction work

Construction work covers many types of projects, and each one will require different types of planning, numbers of building operatives, tools, equipment, materials and processes. The size and location of the project, together with various planning and environmental constraints will also have an impact.

The following should be considered for all building processes:

- **Sustainability** and environmental protection in relation to the design, planning and delivery stages of project development, across different types and scales of construction project. In construction, sustainability is a broad term that describes how buildings can be constructed in a way that has less impact on the environment. For example, the concept of reduce, recycle and reuse makes better use of natural resources and leads to fewer natural resources being used or wasted.

The reuse of bricks, concrete, etc., for hardcore is one example of making construction work more sustainable.

- Local sourcing, resource protection, reuse and refurbishment of materials and waste management should be built into the process from design to completion.

KEY TERM

Sustainability: ensuring the world's natural resources are not used up today, thereby leaving nothing for future generations. Oil and gas reserves are limited, and alternative sources of energy must be found before they are all gone. Trees cut down for wood must be replaced with new trees so that sources of wood do not run out.

The range of constructions projects includes new builds, renovations, maintenance and restoration.

New build

This type of project involves the construction of new buildings using new materials. It is typically associated with domestic housing, but can be other new buildings, such as hospitals, factories and leisure complexes.

▲ Figure 1.2 A new build construction site

Renovation

Renovation is the process of improving or modernising an old building, so it is returned to a good state of repair. The process typically involves replacing old with new.

▲ Figure 1.3 Renovation work in progress

Maintenance

Maintenance refers to the process of ensuring that buildings remain in good order, free from decay, and function as they were intended. The most effective form of maintenance is to have a planned programme throughout the life of a building. This could involve weekly, monthly or annual maintenance routines that can extend the life of the building.

▲ Figure 1.4 Decorating is a form of maintenance

Restoration

Restoration is the process of returning a building to its former state but may also improve it. This type of work is normally undertaken on buildings of historic interest. Restoration involves using materials and processes to authentically match the existing building structures, or as they appeared originally. Some structures of architectural interest or historic significance may be protected as **listed buildings**.

Listing indicates a building's special architectural and historic interest and brings it under the consideration of the planning system, so that it can be protected for future generations. The older a building is, and the fewer the surviving examples of its kind, the more likely it is to be listed. Listed buildings are graded as follows:

- Grade I buildings are of exceptional interest, only 2.5 per cent of listed buildings are Grade I.
- Grade II* buildings are particularly important buildings of more than special interest; 5.8 per cent of listed buildings are Grade II*.
- Grade II buildings are of special interest; 91.7 per cent of all listed buildings are in this class and it is the most likely grade of listing for a homeowner.

▲ Figure 1.5 You may be required to work on buildings of historical interest

KEY TERM

Listed building: a building that can only be restored, altered or extended if consent is given under government planning guidance. A listed building is given a grade which indicates its level of special interest or significance.

Types of construction work

Construction projects will usually fall into one of three categories – domestic/residential, commercial or industrial.

Domestic or residential

Domestic or residential construction typically refers to buildings that will be occupied as a person's home. This category includes houses, flats, bungalows, terraced properties and other types of residential building that are built for the purpose of living in. This also includes housing of multiple occupancy and high-rise flats.

▲ Figure 1.6 Residential housing

Commercial

Commercial typically refers to all types of land or building premises that are used for business purposes and are non-residential. Commercial properties include:
- schools, colleges and universities that are built for the purpose of providing spaces for education
- health centres, chemists and doctors' surgeries that are built to provide healthcare
- offices, which may be of high-rise or low-level construction and are typically used for business purposes
- entertainment areas, including cinemas, theatres, sports grounds and other buildings used for leisure and entertainment purposes
- light industrial units for small businesses
- vehicle showrooms
- shops including high-street shops and out-of-town stores.

Some commercial properties have a residential element to them, for example hospitals, care homes, prisons, hotels, hostels and student halls of residence. These partly fit within the residential context, but they are likely to be considered commercial projects due to their size.

▲ Figure 1.7 Commercial offices

Industrial

Industrial buildings are those that are used by an industry to manufacture or process articles as well as those concerned with transport infrastructure. Bridges and tunnels may also be classed as industrial structures.

This type of building includes:
- factories, for example vehicle, domestic appliance or food production buildings used for manufacturing processes
- quarry processing plants for cement, sand, gravel or other building products that have been extracted from the ground
- factories where steel and timber are produced, which among other uses may be used in the construction of buildings
- power stations using gas, electricity, oil or nuclear fuel, as well as wind farms and other constructions used for this purpose
- petrol stations
- bridges and tunnels – structures that provide access from one part of land to another and may involve tunnelling under or building over land or water, for example the Channel Tunnel and the Dartford Crossing Bridge
- rail and other transport constructions, such as railway stations, bus stations, airports and other buildings associated with travel.

▲ Figure 1.8 Industrial building

Organisations that contribute to the construction process

Building contractors

A building contractor is a person or firm that undertakes a contract to provide materials and/or labour to perform a service or do a job. In the case of the construction industry, building contractors carry out construction work and erect buildings. Building companies may be small, medium or large and comprise the principal or main contractor and various subcontractors. Small companies may range from a single person (sometimes referred to as a jobbing builder) up to those employing 50 workers. A medium-sized contractor typically directly employs between 50 and 250 employees, and a large contractor will directly employ over 250 people.

Manufacturers

A manufacturer is a person or a registered company which makes finished products from raw materials. Manufacturers produce the materials and equipment used in construction, such as bricks, processing timber, plaster, sand and gravel from quarries, paints, toilets, baths, sinks, boilers, electrical sockets, cables and many other items used in the building process.

Suppliers

A supplier is a person or organisation that provides something needed during construction, such as a product or service. Suppliers, or builders' merchants, sell goods from a manufacturer to the end users, such as building contractors, who use them in construction projects.

Local authorities

A local authority is an administrative body in local government that is officially responsible for all the public services and facilities in an area of the country. They ensure that all building and construction projects in their area comply with legislation relating to planning and building control, are in line with the local development plan and conform to building regulations.

Legislative bodies

The government passes legislation that lays down the rights and responsibilities of individuals and authorities in all aspects of life. The government then instructs an institution, department or body to oversee and control the implementation of that law.

The Health and Safety at Work etc. Act 1974 is just one of many laws related to construction. This Act is administered by the Health and Safety Executive (HSE). The HSE is a UK government agency responsible for the encouragement, regulation and enforcement of workplace health, safety and welfare, and for research into occupational risks in the UK.

The Building Regulations (2010) is another example of legislation implemented by the government to enable buildings to be built safely and to certain approved standards. Building control officers are appointed by the local authority and are responsible for ensuring these standards are maintained.

> **ACTIVITY**
>
> Visit the Planning Portal at www.planningportal.co.uk and navigate to the building control section. In your own words describe the main purpose of the Building Regulations 2010 regarding materials and workmanship (further information on this topic can be found in the Approved Documents, Regulation 7).

Roles of construction team members

The construction of a building is a complex process that requires the involvement of many individuals. They make up the construction team and it is important that they work well together. Figure 1.9 shows a typical team involved in a large construction project.

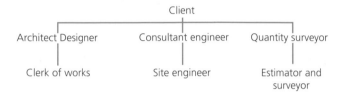

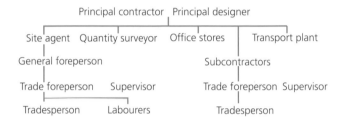

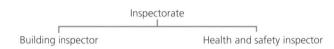

▲ Figure 1.9 Members of the construction team and their roles

The members of the construction team are generally classified as professional, craft or operative.

Professional

Those members of the team classified as professional will generally have a management role, where they have a specific responsibility for the planning, design or management of the construction project. The table below describes some of the key roles of the professional team. Members of the legislative bodies, such as planners and inspectors, are also described as carrying out a professional role. There will also be technicians and many other support staff working with the various professional roles listed, as well as several specialist roles carried out in the construction process. For example, an architectural technician will typically work in support of the architect to carry out various duties in the construction design process. The Principal contractor and Principal designer will ensure that all aspects related to the Construction Design and Management Regulations are formulated and implemented. The full details of the CDM regulations are explained on page 9 and 26. They may appoint a single co-ordinator on contracts to oversee this role.

▲ Figure 1.10 Members of the professional construction team

Architect	Designs new buildings and the spaces around them; he or she also works on the restoration and conservation of existing buildings
Engineer	Civil engineer: designs and manages construction projects ranging from bridges and buildings to transport links and sports stadiums
	Building services engineer: designs, installs and services equipment and systems in buildings like offices and shops
	Structural engineer: helps to design and build large structures and buildings, such as hospitals, sports stadiums and bridges
Designer	May be involved in interiors or landscaping and is generally involved in the **aesthetic** layout of spaces. This could include furnishings, plants and other works of a decorative nature
Surveyor	Quantity surveyor: oversees construction projects, manages risks and controls costs
	Building surveyor: advises clients about the design, construction, maintenance and repair of buildings
	Land surveyor: measures the shape of the land and gathers data for civil engineering and construction projects

Estimator	Works out how much it will cost for a company to supply products and services to its clients
Site or construction manager	Organises the work on building projects, making sure it is completed safely, within budget and on time
Clerk of works	Oversees the quality and safety of work on a construction site, making sure that building plans and specifications are being followed correctly
Town planner	Helps shape the way towns and cities develop and balances the demands on land with the needs of the community
Building control officer	Makes sure building regulations are followed
Health and safety inspector and CDM co-ordinator	Health and safety inspector: ensures that workplaces are safe for workers and members of the public, and checks that employers and employees follow health and safety regulations Construction design and management co-ordinator: advises the client on health and safety matters during the design process and planning phases of construction

KEY TERM

Aesthetic: the beauty of something and how that is appreciated by the person looking at it.

Craft

The definition of a craftsperson relates to an individual that has been trained and qualified to perform a task or job. The table below highlights some of the key roles carried out by members of the craft team and other skilled and trained workers that apply their specialist skills within a construction project. The management of the day-to-day activities of this skilled group may be undertaken by a supervisor, foreperson or chargehand, many of whom will also be skilled craftspeople who have a practical working role as well.

▲ Figure 1.11 Scaffolders working on site

Bricklayer	Builds houses, repairs walls and chimneys and refurbishes decorative stonework. He or she also works on restoration projects
Carpenter and joiner	Makes and installs wooden structures, fittings and furniture
Dry liner	Uses plasterboard panels to build internal walls, suspended ceilings and raised flooring in houses, offices and shops
Electrician	Installs, services and fixes electrical equipment, circuits, machinery and wiring
Foreperson	Often known as foreman, but this role can be carried out by a person of any gender. The individual takes day-to-day responsibility for managing the work of a team of workers. A **general foreperson** may take responsibility for several trades and a **trade foreperson** may take responsibility for a particular trade, e.g. foreperson bricklayer
Painter and decorator	Prepares and applies paint, wallpaper and finishes to different surfaces
Plant operator	Works with machinery and equipment used on building sites
Plasterer	Prepares and plasters walls and ceilings ready for decoration and finishing
Plumber	Installs and services hot and cold water systems, heating systems and drainage networks
Roofer	Re-slates and tiles roofs, fits skylight windows and replaces lead sheeting and cladding
Scaffolder	Designs, erects and dismantles scaffolding on buildings that allows workers to work safely at height
Stonemason	Carves blocks of stone and lays and fits stonework into place on construction projects
Tiler	Tiles walls and floors in kitchens, bathrooms, shops, hotels and restaurants

Operatives

Sometimes referred to as labourers or construction operatives, these roles involve a range of practical tasks that help skilled construction workers. This could include digging trenches, mixing and laying concrete, operating machinery, using equipment such as cement mixers, drills and pumps, and moving, loading and unloading materials.

▲ Figure 1.12 Building operatives working on site

Construction industry career opportunities

The Technical Certificate in Painting and Decorating is aimed at those looking to work in the construction industry specifically as a craftsperson in painting and decorating. A painter and decorator is an important part of any construction team, as he or she has the skills and knowledge to use a range of coverings, such as paint and wallpaper, to enhance and protect plaster, metal and wood surfaces. The Technical Certificate allows you to gain an understanding of the skills and knowledge that are important when working as a painter and decorator and to progress to further training in this area. This qualification is designed to help you enter employment in the construction industry as a painter and decorator.

Progression routes

- **National Vocation Qualification (NVQ):** this is a work-based qualification (available at various levels) that combines learning and practical working.

- **Apprenticeship:** allows you to gain a qualification (at various levels) while working alongside someone who is already qualified and experienced in their role. An apprenticeship combines practical training in a job with study, usually one day a week.
- **Degree-level course:** these are usually completed at college or university over three to four years. Depending on the subject, these courses combine coursework, exams and practical learning. It is possible to study for a degree in your own time or part-time combined with work and supported by your employer.

> ### ACTIVITY
> Visit www.goconstruct.org/ or https://nationalcareers.service.gov.uk/ and find out more about apprenticeships and the various job roles in construction.

Importance of qualifications and continuing professional development

Once you are qualified, you can apply for a Construction Skills Certification Scheme (CSCS) card. This card provides proof that you can work on a construction site and have the appropriate training and qualifications for the jobs you do on site. The CSCS card ensures the workforce are appropriately qualified, and the scheme plays an important part in improving standards and safety on UK construction sites. Most principal contractors and major house builders require construction workers on their sites to hold a valid CSCS card.

Continuing professional development (CPD) means continuing to update and expand your knowledge and skills, even when qualified, to ensure that you always work in line with current industry practice.

> ### ACTIVITY
> Visit www.cscs.uk.com/ and find out what qualifications are required to apply for a skilled card as a painter and decorator.

Communication within a construction team

Good communication is essential in the day-to-day running of a construction project. For each project there will be a chain of **hierarchy** that enables information to be properly managed and documented. The chain of hierarchy will vary depending on the size of the project. For example, on a large project it is usual for the client to communicate their requirements to the architect who will in turn share the information with the main contractor for further sharing through the chain of command to the workforce.

▲ Figure 1.13 Meeting on site

Key personnel involved in day-to-day communication

The **site manager** is responsible for overseeing the project and communicating the **main contractor's** directions. He or she will communicate with the various **supervisors** and **forepersons** and they will in turn communicate and direct the activities of the individuals in the workforce. Through this chain of communication, the wishes of the client are carried out.

Further daily communication is carried out by the various operatives organising and working together to ensure the project runs in the most efficient manner for all concerned.

Others involved in wider communication

In a large building project, there will be many other individuals and bodies that have a role to play in the wider communication of information. The **architect** will carry out the design to meet the needs of the client. He or she will liaise with local authority planners and environmental and building control officers to ensure the project meets all the legislative requirements before passing information on to the main contractor.

In turn, the main contractor will appoint a **health and safety co-ordinator** to plan how the project will be set up and managed safely **and a construction design and management (CDM) co-ordinator** to help everyone involved in the project to communicate effectively. The CDM co-ordinator's role is to advise the client on matters relating to health and safety during the design process and during the planning phases of construction. The principal duties of the CDM co-ordinator are as follows.

- Notify the Health and Safety Executive of the particulars of a project, as specified in schedule 1 of the Regulations using Form F10. The HSE is informed if a project is likely to last longer than 30 days or involve more than 500 person-days of construction work.
- Advise the client as to whether the materials and processes selected for the project are acceptable in terms of quality and quantity.
- Co-ordinate health and safety aspects of design work and co-operate with others involved with the project to allow the work to proceed in a safe manner.
- Help with communication between the client, designers and contractors.
- Provide, or ensure that the client provides, pre-construction information.

- Advise on the construction phase plan before construction works begin.
- Advise on any subsequent changes to the construction phase plan.
- Work with the principal contractor regarding any ongoing design work during construction.
- Prepare the health and safety file, and then give the health and safety file to the client at the end of the construction phase.

The HSE will appoint inspectors who will carry out inspections of building works to ensure that laws relating to health and safety are being adhered to. The inspector will:

- investigate when an accident occurs, something has gone wrong or a complaint is made, to find out whether people working on site or nearby are at risk
- ensure action is taken to control risks and when someone is breaking the law
- take appropriate enforcement action in relation to any **non-compliance**, ranging from advice on stopping dangerous work activities to prosecuting when people are put at serious risk of harm
- provide advice and guidance to help everyone to comply with the law and avoid injuries and ill health at work.

Inspectors have the right of entry to your premises as well as the right to talk to employees and safety representatives, and to exercise powers necessary to help them fulfil their role.

KEY TERM

Non-compliance: not following the requirements of legislation, for example the Health and Safety at Work Act, and potentially working in an unsafe manner.

Local residents

It is important that the community, and especially owners or residents of adjoining properties, are kept informed about the ongoing construction work, and any disruption to their daily lives should be kept to a minimum. They should be told of disruption to electricity, gas or water supplies or any other issues in a polite, respectful way to reduce potential conflict during the construction process.

Methods of communication

Written

The written form of communication is vital within construction and is used in many ways within a project. In most cases the formal documents that are produced by the architect on behalf of the client will be in a written format. These form part of the contract and will therefore have a legal bearing, particularly in any dispute regarding design, costs and other building processes. The documents that are communicated between clients, architects and contractors will need to be kept throughout the life of the project as they will be required frequently during the building process for costings, quantities, resources, planning purposes, cost control, programming and to obtain tenders. These documents can now be managed using a **building information modelling (BIM)** system.

Sometimes more informal documents are used for communication purposes, such as emails and handwritten notes, although these can be scanned so that a record is kept.

KEY TERM

Building information modelling (BIM): a digital process for creating and managing information about a construction project throughout its lifecycle, from its earliest conception to completion and potentially its eventual demolition.

Advantages of written communication

Written communication is more accurate and precise and is less likely to lead to confusion or misunderstanding. Examples and illustrations can easily be used too, and a permanent record is created that can be referred back to if needed. Written communication is acceptable as a legal document, so for this reason, any important verbal discussions should be written down to provide confirmation of what was said.

Verbal

Communication with colleagues and line management may involve a face-to-face chat or a more formal site meeting. Communication helps to ensure that work is done on time and to the required standard.

Verbal communication is the commonest form, but text messaging and phone calls can be convenient. Emails or letters ensure a record of the conversation is kept. Day-to-day communication between colleagues and the foreperson will mostly be verbal unless there is formal information to be shared such as colour schedules or other aspects of the specification that are written down so there is no misunderstanding or confusion.

Regular communication reduces the chance of misunderstandings and helps to keep things on track. To communicate effectively, you need to be polite, respectful and to speak appropriately with a wide variety of people while maintaining good eye contact. It is important to listen well, present your ideas appropriately, speak clearly and **concisely** and be professional.

KEY TERM

Concisely: providing information in a few words, so that communication is brief, but with enough detail to cover all the points.

2 UNDERSTAND CONSTRUCTION INFORMATION

Building controls and regulations

There are various controls and regulations that need to be followed when planning and carrying out construction projects, and it is useful to know what they are and how they apply to any project that you work on.

Planning permission

Most building projects require planning permission, particularly new builds but also anything from an extension on a house to a new shopping centre. When renovation projects involve the change of use of a building, for example when a public house is converted into a residential home, planning permission is required

for the change to take place. Local authorities are responsible for deciding whether a building project should go ahead.

Building regulations

Building regulations are minimum standards required by law for the design, construction and alteration of almost all building work. The regulations were developed by the UK government and approved by parliament. The Building Regulations 2010 cover the construction and extension of buildings and these regulations are supported by Approved Documents. Approved Documents set out detailed practical guidance on compliance with the regulations. Building regulations approval is different from planning permission, and both may be needed for a project.

An application for building regulations approval can be made to any local authority building control department or approved inspector. The British government has developed a one-stop shop website for planning and building approval requirements, called the Planning Portal, www.planningportal.co.uk, which has lots of information on this topic.

Health and safety law

Most issues related to health and safety law are covered under the Health and Safety at Work Act (HSWA) 1974. This Act places a legal duty on employers to ensure, so far as reasonably practicable, the health, safety and welfare of employees, and to ensure that employees and others are kept safe. Under the Act, employers have an obligation to ensure any potential risk of work-related injury is eliminated or controlled. If there are five or more employees then there must be a written health and safety policy statement, setting out how the management of health and safety is covered in the organisation.

The management of health and safety is further covered under the Management of Health and Safety at Work Regulations (MHSWR) 1999. These Regulations require employers to consider the health and safety risks to employees and to carry out a **risk assessment** to protect employees from exposure to reasonably foreseeable **hazards** and risks. Those risks include work-related violence.

The Construction Design Management Regulations (CDM 2015) are the main regulations for managing the health, safety and welfare of those working on construction projects. CDM Regulations apply to all building and construction work, including new build, demolition, refurbishment, extensions, conversions, repair and maintenance.

Further information about the Health and Safety at Work Act and other health and safety law related to construction is provided later in this chapter.

Quality and standards

Most industry sectors have a set of standards, and there are currently over 3500 standards relevant to construction in the UK. Meeting a high standard of quality, whether in the supply or use of materials, is a key issue for the construction industry. Quality of workmanship in all areas is also important. Most standards are set out in a series of documents supplied and managed by the British Standards Institute (BSI). British Standard (BS) publications are technical specifications or practices that can be used as guidance for the production of a product, for carrying out a process or for providing a service. The BSI Kitemark, first introduced in 1903, is commonly found on a range of products, including construction products.

▲ Figure 1.14 The BSI Kitemark confirms that the product that carries it conforms to the relevant British Standards

Quality Management ISO 9001 is the international standard covering a range of quality management standards, including customer focus, managing the business by a series of processes (rather than the people) and focusing on continual improvement. If used by an employer or on a project, ISO 9001 provides the basis for an overall focus on quality and producing products and projects to a high standard.

Types and uses of construction information

Construction projects are complex and involve a significant amount of information that is used to manage, support and organise the project, as explained below.

Specifications

Specifications are normally drawn up as a written document based on a client's needs. The document sets out the requirements in a logical form and is one part of a **building contract**. The specification will describe the types of materials and finish required and should be read in conjunction with the drawings to ensure that everything follows the correct criteria.

Drawings

Information contained in drawings is an important part of most construction projects. A series of drawings will be produced by the architect or engineer to provide a visual illustration of the project. Being able to view the construction project as a drawing makes it easier to understand than just a written description of the project. A drawing will show the location of the project, layouts, and will include detailed drawings that show how something should be built. It is important that the drawings are read in conjunction with any specification, as the written and drawn detail together are what a construction team need to follow. You will look at some commonly used drawings later in this chapter.

Schedules

Schedules give details of elements of the construction, such as repeated design elements, including doors and windows. A typical door schedule is shown below.

Master internal door schedule							
Ref:	Door size	SO width	SO height	Lintel type	FD30	Self closing	Floor level
D1	838 x 1981	900	2040	BOX	Yes	Yes	GROUND FLOOR
D2	838 x 1981	900	2040	BOX	Yes	Yes	GROUND FLOOR
D3	762 x 1981	824	2040	BOX	No	No	GROUND FLOOR
D4	838 x 1981	900	2040	N/A	Yes	No	GROUND FLOOR
D5	838 x 1981	900	2040	BOX	Yes	Yes	GROUND FLOOR
D6	762 x 1981	824	2040	BOX	Yes	Yes	FIRST FLOOR
D7	762 x 1981	824	2040	BOX	Yes	Yes	FIRST FLOOR
D8	762 x 1981	824	2040	N/A	Yes	No	FIRST FLOOR
D9	762 x 1981	824	2040	BOX	Yes	Yes	FIRST FLOOR
D10	762 x 1981	824	2040	N/A	No	No	FIRST FLOOR
D11	686 x 1981	748	2040	N/A	Yes	No	SECOND FLOOR
D12	762 x 1981	824	2040	BOX	Yes	Yes	SECOND FLOOR
D13	762 x 1981	824	2040	100 HD BOX	Yes	Yes	SECOND FLOOR
D14	686 x 1981	748	2040	N/A	No	No	SECOND FLOOR

Bill of quantities

A bill of quantities is drawn up by the quantity surveyor and describes everything that is contained within the specification, drawings and schedules. A bill of quantities contains general information, including the name of the client, address and details of the site and details of the quality of the materials and workmanship.

BILL OF QUANTITIES						
(Assuming Civil Engineering Standard Method of Measurement (CESSM3) is used.)						
					Amount	
Number	Item description	Unit	Quantity	Rate	£	P
	CLASS A: GENERAL ITEMS Specified Requirements Testing of Materials					
A250	Testing of recycled and secondary aggregates	sum				
	Information to be provided by the contractor					
A290	Production of Materials Management Plan	sum				

Number	Item description	Unit	Quantity	Rate	Amount £	P
	Method Related Charges					
	Recycling Plant/Equipment					
A339.01	Mobilise; Fixed	sum				
A339.02	Operate; Time-related	sum				
A339.03	De-mobilise; Fixed	sum				
	CLASS D: DEMOLITION AND SITE CLEARANCE					
	Other Structures					
D522.01	Other structures; Concrete	sum				
D522.02	Grading/processing of demolition material to produce recycled and secondary aggregates	m³	70			
D522.03	Disposal of demolition material offsite	m³	30			
	CLASS E: EARTHWORKS					
	Excavation Ancillaries					

A bill of quantities will also include descriptions of items in terms of quantities and size and will have space for contractors to estimate the cost of a single item and record their projected costs. The example illustrates just one page from a project; in reality a bill of quantities will have many pages.

Deciding who will do the construction work is an important part of any project. Before the contract can begin, the specification, drawings, schedules and bill of quantities will be sent to a number of contractors (usually a minimum of three) asking them to estimate the cost of the work. Once the contractors have returned their estimates the architect will discuss with the client and make a recommendation of the contractor most suitable to carry out the work. This recommendation is likely to be based on the contractor's experience and availability as well as cost.

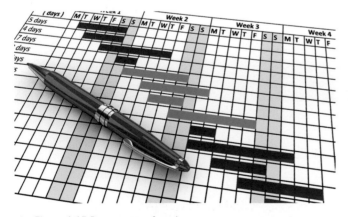

▲ Figure 1.15 Programme of works

Programme of works

Once the contract has been awarded to the chosen contractor and signed, the contractor will put together a programme of works. This describes when various tasks will happen, and when trades, materials and equipment are required. The programme of works is plotted on a chart (often referred to as a Gantt chart) so it can be easily seen when plant and labour are required, and whether each part of the project is running on time.

Building information modelling

Building information modelling (BIM) is an intelligent 3D model-based process that enables engineering, architecture and construction professionals to plan, design, construct and manage a project more efficiently and helps all team members to work to the same standards as one another. Typically, BIM is used on large construction projects, especially those awarded by government. It can be continually updated so that if amendments are made, they are automatically saved and shared with all members of the project team. All the documents listed above will form part of the information sharing, along with communications notes and all the documentation connected with the project throughout its **lifecycle**.

KEY TERM

Lifecycle: the entire time that something exists. For a building, this includes its design, construction, operation and eventual disposal.

Technical drawings used in the construction industry

Each part of a construction project has to be designed and the details of each design communicated to the project team who will build each component and put components together. One way of communicating the designs and details is to produce drawings that represent how the building will look and how it is put together.

Drawing methods

Manual drafting

Manual drafting involves using just hands and pen or pencil to draw lines and shapes to produce the drawings used in a construction project. Typically, tracing paper, a drawing board, T-square for horizontal and vertical lines, set squares, scale ruler and pens and pencils are used to produce the drawings. The tracing paper can then be copied to enable multiple copies to be shared with those involved in the project.

Figure 1.17 shows a T square, set squares, protractor, French curves and a graph template, which are all items that can be used in manual drafting techniques.

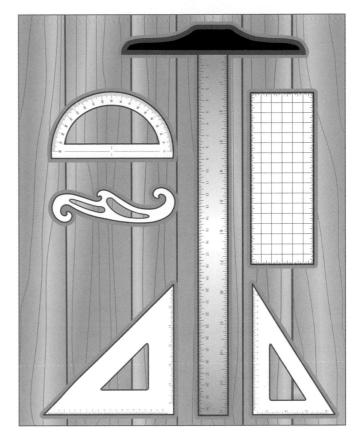

▲ Figure 1.17 Equipment used in manual drawings

CAD and CAD/CAM

▲ Figure 1.18 Computer-aided design

CAD (computer-aided design) refers to the use of software to create precise technical drawings or technical illustrations when designing a product or object as part of the construction process. CAD can be used to create 2D technical drawings or 3D models of items and can also be used to generate animations and other presentational material. Dimensions, descriptions of components and references to specifications, etc.,

▲ Figure 1.16 Manual drawing

can be added so that the CAD diagram is as useful as possible to the project team.

CAD/CAM (computer-aided design/computer-aided manufacture) is used to design items and produce computer programs that can operate machinery to manufacture multiple items in a factory or workshop. Robots can be programmed to manufacture parts and carry out tasks that enable the construction of buildings and their elements within a factory ready for delivery to site.

Drawing information

Buildings and their components need to be scaled down, so they fit onto sheets of paper when drawn. Drawings show the positions of rooms, windows, doors, kitchen units and so on. When elevations are shown, more detail can be added with measurements related to the vertical view.

Some of the **scales** commonly used for elevations are 1:200, 1:100, 1:50, 1:10, 1:5 and 1:1. A scale of 1:1 is full size.

KEY TERM

Scale: the relation between the real size of something and its size on a drawing. For example, 1:1 is full size, while 1:5 means the drawing is one-fifth of the size of the real object.

IMPROVE YOUR MATHS

Using a tape measure, measure a wall or rectangular object within the room you are in. Take note of the measurements in either metres or millimetres. What size would the object be drawn if it was to be drawn at a 1:5 representation?

For example, a wall that is 2400 mm wide and 1200 mm high (2400 mm × 1200 mm) would be drawn at 480 mm × 240 mm at a scale of 1:5.

Types of drawing

Drawings are used in construction projects to provide details and information. The types of drawings that conform to a set of common standards which are accepted and used by the construction industry are described as follows.

Orthographic projection

Orthographic projection is a drawing technique used to represent three-dimensional objects as a series of two-dimensional 'flat' drawings in which there is no perspective. Orthographic projection is a type of 'parallel' projection in which the four views of an object are shown. The orthographic projection commonly used in the UK is called first angle projection. In Figure 1.19, all four sides and the plan view of the building are shown.

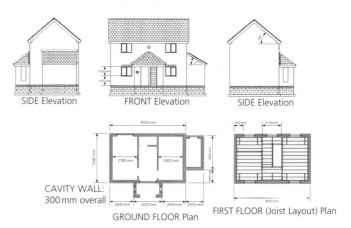

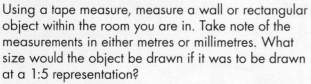

▲ Figure 1.19 Orthographic projection

Isometric projection

Isometric projections are a pictorial projection of a solid object on a flat surface. They are drawn so that all vertical lines remain vertical and their length is to scale, but all horizontal lines are drawn at an angle of 30°. They should not be used for scale drawings but just used to provide a three-dimensional view.

▲ Figure 1.20 A scale rule that has pre-marked scales on the side is used to ensure drawings are accurate

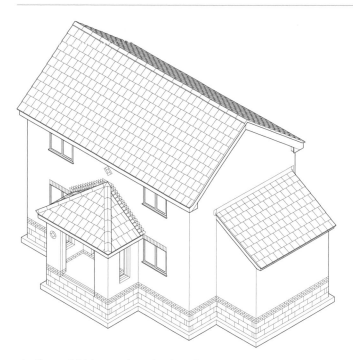

▲ Figure 1.21 Isometric projection of a house with porch and side extension

Location drawings

These are usually prepared by an architect or architectural technician. They show the location of the building plot, position of the building and areas within the building.

Block or location plans

A block plan shows a property in relation to the surrounding properties. The scale used tends to be 1:1250 or 1:2500. Very little detail is available from this type of plan. The direction north is usually shown.

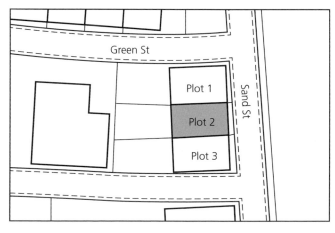

▲ Figure 1.22 Block plan

Site plans

Site plans show the plot in more detail, with drainage and road layouts. In Figure 1.23, the size and position of the existing building and the proposed extensions are shown in relation to the boundary of the property. A scale of 1:200 or 1:500 is usually used.

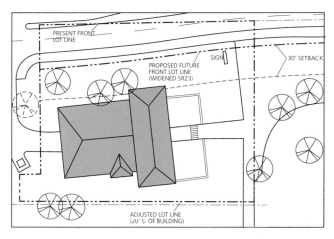

▲ Figure 1.23 Site plan

Floor plans

Floor plans show the position of walls and size of rooms along with the layout of elements within the building, such as kitchen units and bathroom suites. A scale of 1:100 or 1:50 is commonly used for floor plans.

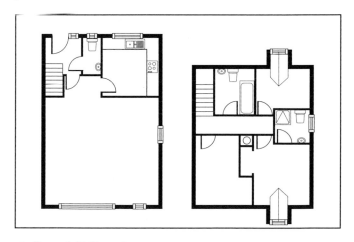

▲ Figure 1.24 Floor plan

Elevations

An elevation shows a particular face of a building. Figure 1.25 is an exterior elevation that shows the roof, doors and windows. Interior elevation views are also used to show the vertical layout of the walls, which is useful when planning kitchens and other fitted furniture. Scales of 1:100, 1.50 and 1:20 are commonly used for elevations.

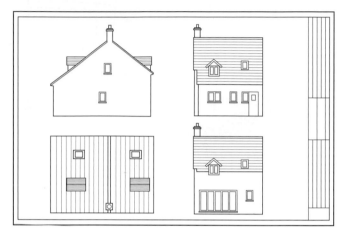

▲ Figure 1.25 Elevations

Sections

A section drawing is a cut through of a part of a building or a component to show greater detail. The section drawing in Figure 1.26 shows a cut through of the building and its construction from foundation level through to roof level. Scales of 1:100, 1:50 and 1:20 are commonly used for section drawings.

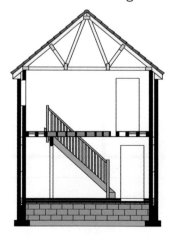

▲ Figure 1.26 Section

Detail, assembly or component drawings

Detail, assembly or component drawings show more detail and sometimes include written instructions. The drawing in Figure 1.27 shows details of the construction of a cupboard. Scales of 1:10, 1:5, 1:2 and 1:1 are normally used to show how components are assembled.

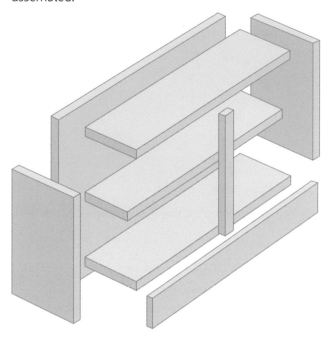

▲ Figure 1.27 Assembly drawing

Symbols and hatchings

Symbols and hatching are used on drawings to indicate different types of material or elements of the building. The same symbols are used for all construction drawings, whether manually drafted or using CAD. These common symbols and hatchings conform to a British Standard format and were developed to allow all users to recognise standardised representations used within construction drawings. The British Standard used to enable a common understanding of the information in drawings is BS 8541-5:2015 Library objects for architecture, engineering and construction.

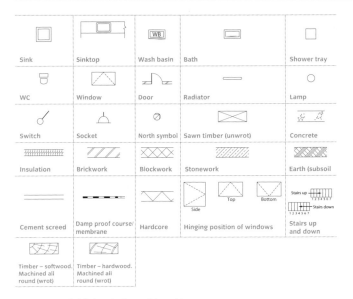

▲ Figure 1.28 Symbols and hatchings

Health and safety and construction sites

This section provides an overview of the relevant legislation and guidance on health, safety and welfare in relation to working on a construction site.

Reference is made throughout this section to the Health and Safety Executive (HSE) website (www. HSE.gov.uk), which has concise and easy to follow information, and it is recommended that you become familiar with this resource. Some of the activities in this section require you to access and download information from the HSE website.

The HSE enforces health and safety regulations when its inspectors visit construction sites to provide advice and guidance. If the site is engaging in unsafe activities an inspector may issue an enforcement or prohibition.

A good understanding and awareness of health and safety issues are required when working in construction, but at Level 2 you do not need to know all the detail. Figure 1.29 shows some of the health and safety requirements that are commonly found on construction sites.

▲ Figure 1.29 Safety signs

Health and Safety at Work etc. Act 1974

We all have a right to work in places where risks to our health and safety are properly controlled. The Health and Safety at Work etc. Act 1974 applies to all workplaces, and everyone who works on a building site or in a workshop is covered by this legislation. This includes employed and self-employed operatives, subcontractors, the employer and those delivering goods to the site. The Act not only protects those working on a construction site, it also ensures the safety of anyone else who might be nearby. The Health and Safety at Work etc. Act is sometimes referred to as HSWA or HASAWA.

Health and safety is about stopping you getting hurt at work or ill through work. Your employer is responsible for health and safety, but you have a role to play as well.

Key employer responsibilities

The key employer health and safety responsibilities under HASAWA are to:

- provide a safe working environment
- provide safe access (entrance) and egress (exit) to the work area
- provide adequate staff training
- have a written health and safety policy in place
- provide health and safety information and display the appropriate signs
- carry out risk assessments
- provide safe machinery and equipment and ensure it is well maintained and in a safe condition
- provide adequate supervision to ensure safe practices are carried out
- involve trade union safety representatives, where appointed, in matters relating to health and safety
- provide personal protective equipment (PPE) free of charge
- ensure the appropriate PPE is used whenever needed, and that operatives are properly supervised
- ensure materials and substances are transported, used and stored safely.

Key employee responsibilities

- Follow the training you have received when using any work items your employer has given you.
- Take reasonable care of your own and other people's health and safety.
- Co-operate with your employer on health and safety and wear the PPE supplied.
- Tell someone (your employer, supervisor or health and safety representative) if you think the work or inadequate precautions are putting anyone's health and safety at serious risk.
- Do not use drugs, medication or alcohol when working on a construction site as this can lead to poor judgement which could affect performance and lead to an increased risk of accidents.

Site induction

When you arrive on a new site you should be given an induction. Inductions make sure you know the key points related to your health, safety and welfare when on site. It is the responsibility of your employer to arrange the induction to take place before you begin work.

It is likely that your employer will make arrangements for the induction to be delivered by the site manager or possibly the site foreperson, who will follow the HSE's Guidance on Regulations, Managing Health and Safety in Construction: Construction (Design and Management) Regulations 2015. The Guidance on Regulations covers the following:

- senior management commitment to health and safety
- outline of the project
- management of the project
- first aid arrangements
- accident and incident reporting arrangements
- arrangements for briefing workers on an ongoing basis, for example toolbox talks
- arrangements for consulting the workforce on health and safety matters
- individual worker's responsibility for health and safety.

The induction might include information about:

- site layout
- site welfare facilities
- site rules
- details of key personnel such as supervisors, safety officers, first aiders and fire marshals
- responsibilities and legal duties
- co-ordination between contractors
- fire safety
- emergency procedures and muster points
- location of first aid equipment and fire extinguishers
- permit to work systems
- procedures for signing in and out
- parking arrangements
- method statements
- tools, plant, vehicles and equipment
- operating hours
- cleanliness and waste management
- meaning of signs

- smoking, drugs and alcohol
- standard of behaviour expected
- personal protective equipment
- working at height
- ear defender zones
- restricted areas
- handling goods and materials
- asbestos
- vibration
- accident reporting.

▲ Figure 1.30 Site induction meeting

Toolbox talks

A 'toolbox talk' is a short presentation to the workforce on a single aspect of health and safety. It is quite common for activities to change on a daily basis on site, and a toolbox talk is a way of updating site personnel on some of the important aspects that might affect their safety. It is common for there to be a large board of daily activities relating to such things as the use of cranes, lifting equipment and access arrangements on a worksite.

Safety signs

There are a number of signs that you will encounter on a construction site, and they fall into four main categories:

- things you must not do (red and white prohibition signs)
- things you must do (blue mandatory signs)
- warnings to be careful of potential risk or danger (yellow and black hazard or caution signs)
- directions to safety or in the case of emergencies (green safe condition signs).

Type	Description
Prohibition	A sign prohibiting behaviour that is likely to increase or cause danger, e.g. 'No access for unauthorised persons'
Mandatory	A sign requiring specific behaviour, e.g. 'Eye protection must be worn'

Type	Description
Caution	A sign giving warning of a hazard or danger, e.g. 'Danger: slip hazard'
Safe condition	A sign giving information related to emergency exits, first aid or rescue facilities, e.g. 'Fire assembly point'

ACTIVITY

Figure 1.31 Safety signs

Name each of the signs shown in the illustration. Use the HSE website to search for the Health and Safety (Safety Signs and Signals) Regulations 1996, Guidance on Regulations document L64, to help you. You may wish to number each sign, so your answers are clear, for example top row 1 to 7, etc.

Fire safety

An awareness of fire safety and how to avoid the spread of fire on a building site is essential. It is important to keep work areas tidy, with timber offcuts, paper and cardboard cleared away and any flammable material or liquid removed to the appropriate areas, either for recycling or removal. Ensure that smoking and naked flames are in controlled spaces or permitted areas.

Figure 1.32 shows the three things needed for fire to start: oxygen, heat and fuel.

1 **Oxygen**: a gas in the air that combines with flammable substances under certain circumstances.
2 **Heat**: a source of fire, such as a hot spark from a grinder or naked flame.
3 **Fuel**: things that will burn, such as flammable solvents, timber, cardboard or paper.

If one of these is missing, there will be no fire. If all are present, then a fire will occur. If any of the three things are removed or missing, then the fire will go out.

By being careful with heat and tidying away potential sources of fuel you reduce the risk of fires starting and spreading.

▲ Figure 1.33 Fire blanket

▲ Figure 1.32 The fire triangle

Fire extinguishers

It may be possible to control the spread of flame and fire with the use of a fire blanket (Figure 1.33) or fire extinguishers (Figure 1.34). The correct extinguisher and method to deal with a fire depends on the type of fire and its severity. It is important that you know where fire extinguishers and fire blankets are located, and which fire extinguishers can be used on different fires. The table to the right shows which fire extinguisher is used for each type of fire.

Should the fire be too large or difficult to put out then the fire service should be called. Fires on building sites should in the first instance be reported to the supervisor/foreperson, and then the site manager will be contacted.

▲ Figure 1.34 Types and uses of fire extinguishers

Class of fire	Materials	Type of extinguisher
A	Wood, paper, hair, textiles	Water, foam, dry powder, wet chemical
B	Flammable liquids	Foam, dry powder, CO_2
C	Flammable gases	Dry powder, CO_2
D	Flammable metals	Specially formulated dry powder
E	Electrical fires	CO_2, dry powder
F	Cooking oils	Wet chemical, fire blanket

Reporting of Injuries, Diseases and Dangerous Occurrences Regulations (RIDDOR) 2013

▲ Figure 1.35 Injury report form

Fatal and specified non-fatal injuries to workers and members of the public are reported by employers using guidelines from the Reporting of Injuries, Diseases and Dangerous Occurrences Regulations (**RIDDOR**) 2013. Figure 1.35 shows an example of the reporting form that needs to be completed to report anything described under this Regulation.

RIDDOR Regulations 4–6 cover the reporting of work-related deaths and injuries other than for certain gas incidents. Deaths and injuries need to be reported when:

- there has been an accident which caused the injury
- the accident was work-related
- the injury is of a type which is reportable.

It is important to know just what RIDDOR means when it uses terms such as 'accident'. Here are some key terms and their definitions from HSE documentation.

- **Accident**: a separate, identifiable, unintended incident, which causes physical injury. This specifically includes acts of violence to people at work. Injuries in themselves, such as feeling a sharp twinge in your back when working, are not accidents. There must be an identifiable external event that causes the injury, such as a falling object striking someone. Repeated exposure to hazards

that eventually cause injury, such as repetitive lifting, are not classed as 'accidents' under RIDDOR.

- **Work-related**: RIDDOR only requires accidents to be reported if they happen in connection with work. The fact that there is an accident at a work premises does not, in itself, mean that the accident is work-related – the work activity itself must contribute to the accident. An accident is 'work-related' if any of the following played a significant role:
  - the way the work was carried out
  - any machinery, plant, substances or equipment used for the work
  - the condition of the site or premises where the accident happened.
- **Reportable injuries**: the following are reportable under RIDDOR when they result from a work-related accident:
  - the death of any person (Regulation 6)
  - specified injuries to workers (Regulation 4)
  - injuries to workers which result in their incapacitation for more than seven days (Regulation 4)
  - injuries to non-workers which result in them being taken directly to hospital for treatment, or specified injuries to non-workers which occur on hospital premises (Regulation 5).

IMPROVE YOUR ENGLISH

Visit the RIDDOR section of the HSE website (www.hse.gov.uk/riddor/specified-injuries.htm) and copy the list of specified injuries to workers that must be reported under RIDDOR.

Control of Substances Hazardous to Health (COSHH)

Every year thousands of workers are made ill by hazardous substances that result in them contracting a lung disease such as asthma, cancer or a skin disease, such as dermatitis.

Control of Substances Hazardous to Health (**COSHH**) **is** designed to control the risks and hazards associated with use of hazardous materials. Measures must be taken to control substances that are hazardous to health. In order of priority, these are:

1 Eliminate the use of a harmful product or substance and use a safer one.
2 Use a safer form of the product, for example paste rather than powder.
3 Change the process to emit less of the substance.
4 Enclose the process so that the product does not escape.
5 Extract emissions of the substance near the source.
6 Have as few workers in harm's way as possible.
7 Provide personal protective equipment (PPE) such as gloves, coveralls and a respirator. PPE must fit the wearer.

If your control measures include 5, 6 and 7, make sure they all work together.

▲ Figure 1.36 Hazardous materials sign

The following substances and situations can be harmful:
● Dusty or fume-laden air can cause lung diseases, particularly in those working as welders, quarry workers or woodworkers.
● Metalworking fluids can grow bacteria and fungi which can cause dermatitis and asthma.
● Flowers, bulbs, fruit and vegetables can cause dermatitis.
● Wet working such as cleaning can cause dermatitis.
● Prolonged contact with wet cement in construction can lead to chemical burns and/or dermatitis.
● Benzene in crude oil can cause leukaemia (a type of cancer).

Many other products or substances used at work can be harmful, such as paint, ink, glue, lubricant, detergent and beauty products. Ill health caused by these

substances is preventable. Many substances can harm health but, used properly, they almost never do. Always check the safety data sheet (see Figure 1.37) that comes with the product.

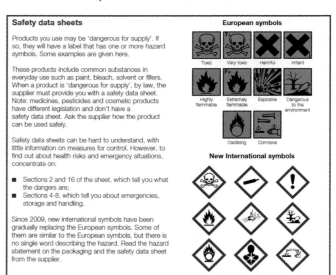

▲ Figure 1.37 Safety data sheet

ACTIVITY

Visit the HSE website and download the leaflet 'Working with substances hazardous to health: A brief guide to COSHH': www.hse.gov.uk/pubns/indg136.pdf

Select a tin of oil-based paint from your workplace or training centre and complete the hazard checklist below. You may also need to visit the manufacturer's website to check all the safety data.

Hazard checklist	Response
Does any product you use have a danger label?	
Does your process produce gas, fume, dust, mist or vapour?	
Is the substance harmful to breathe in?	
Can the substance harm your skin?	
Is it likely that harm could arise because of the way you use or produce it?	
What are you going to do about it? Can you: ● use something else ● use it in another, safer way ● control it to stop harm being caused?	

Construction (Design and Management) Regulations 2015

The Construction (Design and Management) Regulations 2015 (CDM 2015) is the law that applies to the whole construction process on all construction projects, from conception to completion. It outlines what each **dutyholder** must or should do to comply with the law to ensure projects are carried out in a way that keeps workers healthy and safe. All those deemed to be **competent persons** on a construction site will be dutyholders. This means that almost everyone involved in a construction project has legal duties under CDM 2015. The main 'dutyholders' are explained in the table below.

KEY TERMS

Dutyholder: a competent person that has a duty to provide and maintain a work environment without risks to health and safety. Responsibilities include providing and maintaining safe plant, equipment and structures, safe systems of work and the safe use, handling and storage of plant, structures and substances.

Competent person: someone with sufficient training, experience, knowledge and other qualities that allow them to carry out an activity. The level of competence required will depend on the complexity of the situation and in some cases specific training and certification will be required before that person is judged competent.

CDM dutyholder	Summary of role/main duties
Clients	Clients are organisations or individuals for whom a construction project is carried out. The client has to make suitable arrangements for managing a project. This includes making sure: ● other dutyholders are appointed ● sufficient time and resources are allocated to the construction project ● relevant information is prepared and provided to other dutyholders ● the principal designer and principal contractor carry out their duties ● welfare facilities are provided
Domestic clients	Domestic clients are people who have construction work carried out on their own home or the home of a family member that is not done as part of a business. Domestic clients are included in CDM 2015, but their duties as a client are normally transferred to the contractor, on a single contractor project, or the principal contractor on a project involving more than one contractor. However, the domestic client can choose to have a written agreement with the principal designer to carry out the client duties
Designers	Designers prepare or modify designs for a building, product or system relating to construction work. When preparing or modifying designs, they have a duty to eliminate, reduce or control foreseeable risks that may arise during construction and the maintenance and use of a building once it is built. Designers must provide information to other members of the project team to help them fulfil their duties
Principal designers	Principal designers are appointed by the client in projects involving more than one contractor. They can be an organisation or an individual with sufficient knowledge, experience and ability to carry out the role. They plan, manage, monitor and co-ordinate health and safety in the pre-construction phase of a project. This includes: ● identifying, eliminating or controlling foreseeable risks and ensuring designers carry out their duties ● preparing and providing relevant information to other dutyholders and the principal contractor to help them plan, manage, monitor and co-ordinate health and safety in the construction phase
Principal contractor	A principal contractor is appointed by the client to co-ordinate the construction phase of a project if it involves more than one contractor. They plan, manage, monitor and co-ordinate health and safety in the construction phase. This includes: ● liaising with the client and principal designer ● preparing the construction phase plan ● organising co-operation between contractors and co-ordinating their work ● making sure suitable site inductions are provided ● ensuring that reasonable steps are taken to prevent unauthorised access ● making sure workers are consulted and engaged in securing their health and safety ● making sure welfare facilities are provided

CDM dutyholder	Summary of role/main duties
Contractor	A contractor is any individual or business in charge of carrying out construction work (e.g. building, altering, maintaining or demolishing). Anyone who manages this work or directly employs or engages construction workers is a contractor. Their main duty is to plan, manage and monitor the work under their control in a way that ensures the health and safety of anyone it might affect (including members of the public). Contractors work under the control of the principal contractor on projects with more than one contractor
Worker	A worker is an individual who carries out the work involved in building, altering, maintaining or demolishing buildings or structures. Workers include plumbers, electricians, scaffolders, painters, decorators, steel erectors and labourers, as well as supervisors like forepersons and chargehands. Their duties include co-operating with their employer and other dutyholders and reporting anything they see that might endanger the health and safety of themselves or others. Workers must be consulted on matters affecting their health, safety and welfare

On-site welfare facilities

CDM Regulations require appropriate and adequate welfare facilities at most workplaces. The provision of welfare facilities should be considered at the planning stages of a project to ensure they are appropriately located.

Separate toilets and washing facilities should be provided for males and females. These should be ventilated and cleaned regularly. There should be a supply of hot and cold water with soap and towels provided and where necessary showers should be provided if the nature of the work requires regular showering.

Rest rooms for breaks should be provided that enable workers to prepare, heat and eat food and have access to drinking water. Changing areas, ideally with lockers for storage, should also be provided and ideally sited close to the toilets and washing facilities.

Guidance document L153 on the HSE website gives further information on CDM 2015.

Provision and Use of Work Equipment Regulations 1998

The Provision and Use of Work Equipment Regulations 1998 (often abbreviated to PUWER) outline the duties of people and companies who own, operate or have control over work equipment. PUWER also places responsibilities on businesses and organisations whose employees use work equipment, whether owned by them or not. It also covers those who have control over work equipment (e.g. those hiring out work equipment).

'Work equipment' is almost any equipment used by a worker while at work and can include:

- machines, such as sanding machines, steam strippers and spray equipment
- hand tools such as screwdrivers, knives, shave hooks, brushes and rollers
- lifting equipment such as lift trucks, elevating work platforms, vehicle hoists, lifting slings and bath lifts
- other equipment such as ladders and water pressure cleaners.

It is important to use the right equipment for the job, and many accidents happen because people have not chosen the right equipment for the work to be done. Planning ahead helps ensure that suitable equipment or machinery is available.

▲ Figure 1.38 Sander and goggles

In painting and decorating PUWER applies to all the equipment, tools and **access equipment** that is used or operated at work, and the user has a responsibility to check that it is fit for use.

KEY TERM

Access equipment: equipment used in work at height, such as steps, ladders, podiums, towers, etc.

Working with electricity

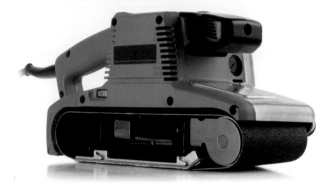

▲ Figure 1.39 Belt sander

Many tools used on a building site need to be powered by electricity. It is important that everyone using electrical tools and equipment has been trained in their use. Tools and equipment also need to be regularly checked to ensure they are safe for use. Electrical tools that are hired should always be checked for an up-to-date Portable Appliance Testing label (PAT testing label).

There are often areas of standing water on construction sites and sometimes cables can become damaged. This has the potential for someone using a piece of electrical equipment to receive an electrical shock, due to the electricity coming into contact with the water through the damaged cable. To reduce the dangers that are associated with using electrical equipment on construction sites reduced-voltage equipment is recommended. Ordinary domestic voltage is 230/240 volts and if there were any faults with the equipment or supply this could potentially cause an electric shock that could kill. In most cases, the lower the voltage the safer it is, so it is recommended to use 110-volt equipment where possible. Transformers are used to step the voltage down from 230/240 V to 110 V.

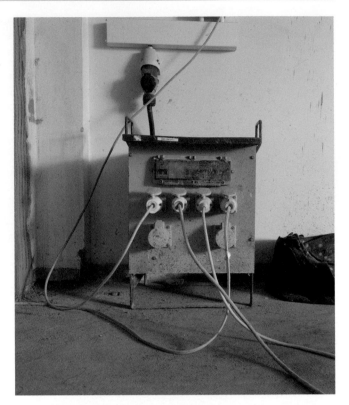

▲ Figure 1.40 A step-down transformer converts voltage to a lower value. Small transformers plug into a wall socket and create low direct current voltages which are used by common electrical tools on a construction site

▲ Figure 1.41 110-volt tools can be recognised by the yellow plugs used

Battery powered tools and equipment are commonly used on site as they are more portable and safer than using mains power. Many different tools are available with voltages varying from 3.6 V up to 36 V masonry drills.

Manual Handling Operations Regulations 1992

Employers must try to avoid their employees undertaking **manual handling**, within reason, if there is a possibility of injury. If manual handling cannot be avoided, then they must reduce the risk of injury by means of a risk assessment.

Incorrect lifting and handling of heavy items can cause damage to your health as you can easily hurt your back. It is important to assess the risk and determine whether it is safe to lift an item on your own. Help may be needed or equipment like a wheelbarrow or some other means of reducing the risk of injury could be used.

It is important to use a correct lifting technique to avoid injury. The safest and most effective way is to use a technique known as **kinetic lifting**. Always lift with your back straight, elbows in, knees bent and feet slightly apart (see Figure 1.42).

▲ Figure 1.42 Safe kinetic lifting technique

KEY TERMS

Manual handling: to transport or support a load (including lifting, putting down, pushing, pulling, carrying or moving) by hand or bodily force.

Kinetic lifting: a method of lifting items where the main force is provided by the operative's own muscular strength. Using the recognised technique will avoid injury.

Personal Protective Equipment at Work Regulations 1992 (as amended)

Employers have duties concerning the provision and use of personal protective equipment (PPE) at work. The HSE website produces a leaflet (INDG174(rev2)) which explains how to meet the requirements of the Personal Protective Equipment at Work Regulations (as amended) 1992.

Personal protective equipment (PPE) protects the user against health and safety risks at work. It includes items such as safety helmets and hard hats, gloves, eye protection, high-visibility clothing, safety footwear and safety harnesses. Note that respiratory protective equipment and hearing protection equipment are covered by other more specific regulations, although any item of PPE provided by an employer must be compatible with this type of equipment.

To make sure the right type of PPE is chosen, you should consider the different hazards in the workplace and identify the PPE that will provide adequate protection against them. This may be different for each job. Figure 1.43 shows the information placed at a site entrance that gives the PPE requirements for that site.

▲ Figure 1.43 Think safety sign

Hard hat or safety helmet 	• A hard hat or safety helmet should be worn to provide protection to the head, primarily from items being dropped or to protect from bumping the head
Safety boots or shoes 	• Safety footwear should be worn to protect the feet from items being dropped, particularly onto the toes • Safety footwear has extra protection around the toe area by the use of steel inserts. The soles of the feet will also be protected from items sticking up on site, such as nails or screws, that could penetrate the sole of ordinary shoes
Ear defenders	• Ear defenders and ear plugs should be worn to provide hearing protection, particularly when carrying out work of a noisy nature • Protection will also be required if working in a noisy environment as working with power tools for long periods can lead to hearing loss or damage
High-visibility (hi vis) vest or jacket 	• High-visibility (or hi vis) clothing is worn to make it easier for other people to see you. This is particularly important when working on site where there may be vehicles or plant working around the area
Safety goggles or glasses 	• Safety goggles or glasses protect the eyes from dust and flying debris while working and should also be worn when working with chemical materials such as paint removers and fungicidal washes
Dust masks and respirators  	• Dust is produced during most construction work and it can be hazardous to the lungs, causing ailments ranging from asthma to cancer. If well fitted, wearing a dust mask helps to filter out dust from the air being breathed • Another hazard is dangerous gases and fumes, such as solvents. A respirator will filter out hazardous fumes, but a dust mask will not! Respirators are rated P1, P2 and P3, with P3 giving the highest protection. If spraying oil-based paints it is advisable to wear a cartridge respirator to provide further filtering of the fine particles of solvent

Gloves	
	● Wearing gloves that are appropriate to the task will protect hands. Decorators may wear light cotton or latex gloves to protect hands from cuts and **abrasions** when rubbing down, for example, and they are comfortable to use
	● Gauntlet gloves should be worn for heavy-duty work such as washing down walls or using liquid paint remover

Sunscreen	
	● When working outdoors in the hot sun the skin can become easily damaged and burning may lead to skin cancer. Exposed skin should be protected using a factor 50 sunscreen
	● It is advised to avoid working outdoors without a shirt on very sunny days

Barrier cream	
	● Barrier cream is an alternative hand protection which will reduce the **absorption** of paint materials into the skin. It provides some protection against irritation and should be rubbed into the hands before carrying out any task using paint
	● Always wash your hands with soap and water after work tasks and before eating food, to avoid **ingesting** any chemicals

Bib and brace overalls or a boiler suit	
	● Overalls with a bib or a boiler suit are useful as protective clothing and enable a small number of tools to be carried in the pockets, making them efficient to use

Safety harness	
	● The full harness shown here is typically worn as part of a fall protection system, which stops a person falling when working high up, such as on bridges or high roof structures. It is made from nylon and a safety lanyard can be attached to it. The lanyard is attached to the safety harness as well as to a secure anchor point and a safety rope
	● The safety rope can be of a fixed length or may incorporate an inertia-operated anchor device that locks to prevent a fall when weight is applied to it. An inertia-operated anchor device attached to a safety line operates in the same way as a car seat belt to stop movement

Respiratory protective equipment (RPE)

▲ Figure 1.44 Respirator

Many workers wear respirators or breathing apparatus to protect their health in the workplace. These devices are collectively known as respiratory protective equipment (RPE). Respirators filter the air to remove harmful substances, and breathing apparatus (BA) provides clean air for the worker to breathe. See the HSE website for guidelines to help you select the correct RPE for the task: 'Respiratory protective equipment at work: A practical guide' (HSG53).

Work at Height Regulations 2005 (as amended)

▲ Figure 1.45 Most building sites will involve work at height

All work at height must be properly planned and organised to ensure safe working practice. The HSE website provides a wealth of knowledge and guidance to help employers and employees ensure they comply with the Work at Height Regulations 2005 (as amended).

The regulations aim to prevent deaths and injuries caused each year by falls at work. They apply to all work at height where it is likely that someone will be injured if they fall. Even a fall from a height of a metre or less could result in an injury, so all risks need to be assessed so that work is planned and carried out safely. Over half of all deaths during work at height involve falls from ladders, scaffolds, working platforms or roof edges or through fragile roofs or roof lights, so always properly assess risks and take a safe approach.

The Work at Height Regulations place duties on employers, the self-employed and any person who controls the work of others (such as facilities managers or building owners who may contract others to work at height).

Those with duties under the regulations must ensure that:

- all work at height is properly planned and organised
- those involved in work at height are competent
- the risks from work at height are assessed and appropriate work equipment is selected and used
- the risks of working on or near fragile surfaces are properly managed
- the equipment used for work at height is properly inspected and maintained.

Can you AVOID working at height in the first place? If NO, got to PREVENT

Do as much work as possible from the ground.

Some practical examples include:

- using extendable tools from ground level to remove the need to climb a ladder
- installing cables at ground level
- lowering a lighting mast to ground level
- ground level assembly of edge protection.

Can you PREVENT a fall from occurring? If NO, go to MINIMISE

You can do this by:

- using an existing place of work that is already safe, e.g. a non-fragile roof with a permanent perimeter guard rail or, if not
- using work equipment to prevent people from falling.

Some practical examples of collective protection when using an existing place of work:

- a concrete flat roof with existing edge protection, or guarded mezzanine floor, or plant or machinery with fixed guard rails around it.

Some practical examples of collective protection using work equipment to prevent a fall:

- mobile elevating work platforms (MEWPs) such as scissor lifts
- tower scaffolds
- scaffolds.

An example of personal protection using work equipment to prevent a fall:

- using a work restraint (travel restriction) system that prevents a worker getting into a fall position.

Can you MINIMISE the distance and/or consequences of a fall?

If the risk of a person falling remains, you must take sufficient measures to minimise the distance and/or consequences of a fall.

Practical examples of collective protection using work equipment to minimise the distance and consequences of a fall:

- safety nets and soft landing systems, e.g. air bags, installed close to the level of the work.

An example of personal protection used to minimise the distance and consequences of a fall:

- industrial rope access, e.g. working on a building façade
- fall-arrest system using a high anchor point.

Using ladders and stepladders

For tasks of low risk and short duration, ladders and stepladders can be a sensible and practical option.

If your risk assessment determines it is correct to use a ladder, you should further MINIMISE the risk by making sure workers:

- use the right type of ladder for the job
- are competent (you can provide adequate training and/or supervision to help)
- use the equipment provided safely and follow a safe system of work
- are fully aware of the risks and measures to help control them.

Follow HSE guidance on safe use of ladders and stepladders at www.hse.gov.uk/work-at-height/index.htm

For each step, consider what is reasonably practicable and use 'collective protection' before 'personal protection'

▲ Figure 1.46 Minimising risks

For managing work at height and selecting the most appropriate equipment, dutyholders must:

- avoid work at height where possible, for example doing the work from ground level using extending equipment
- use work equipment or other measures to prevent falls where work at height cannot be avoided, for example cherry pickers or scaffolding
- use work equipment or other measures to minimise the distance and consequences of potential falls, where the risk cannot be eliminated, for example nets or soft-landing systems.

There is more information on the HSE website – see the Work at Height Regulations 2005 (as amended) and 'Working at height: A brief guide'.

Decorators tend to have to work at height and sometimes in difficult to reach areas, such as over conservatories. A safe device to use in such situations is a mobile elevated work platform called a cherry picker, which can be raised and lowered for access to hard to reach areas at height. (Its name comes from the fact that it can be used to pick cherries or other fruit from trees.)

Common risks associated with working at height can include falling people, tools, materials and equipment, and suitable protection methods should be employed to minimise the risks. This may include guard rails, toe boards, safety nets and fall arrest equipment.

ACTIVITY

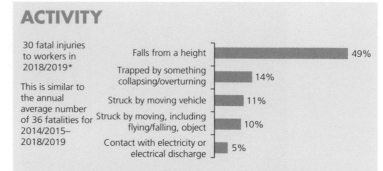

30 fatal injuries to workers in 2018/2019*

This is similar to the annual average number of 36 fatalities for 2014/2015–2018/2019

Falls from a height	49%
Trapped by something collapsing/overturning	14%
Struck by moving vehicle	11%
Struck by moving, including flying/falling, object	10%
Contact with electricity or electrical discharge	5%

Figure 1.47 Fatal accident statistics (Source: RIDDOR 2018/19)

Figure 1.47 is taken from HSE/RIDDOR statistics and shows the numbers of fatal accidents in the construction industry in the UK. Use this information to create a **pie chart** presentation that shows this information in a more colourful way (see Figure 1.48). This can either be drawn by hand or created using a spreadsheet program.

£3.0 billion

£3.4 billion

£8.6 billion

▲ Figure 1.48 Pie chart

KEY TERM

Pie chart: a circle divided into sectors with each slice representing a proportion of the whole.

▲ Figure 1.49 Working at height with guard rails and toe boards at each level

Control of Noise at Work Regulations 2005

The aim of the Control of Noise at Work Regulations is to ensure that workers' hearing is protected from excessive noise that could cause them to lose their hearing and/or to suffer from tinnitus.

Noise at work can cause hearing damage that is permanent and disabling. This can be gradual as a result of exposure to noise over time, but damage can also be caused by sudden, extremely loud, noises. The damage is disabling in that it can stop people being able to understand speech, keep up with conversations or use the telephone. As well as hearing loss, workers may develop tinnitus (ringing, whistling, buzzing or humming in the ears), which is a distressing condition that can also lead to disturbed sleep.

▲ Figure 1.50 Ear defenders

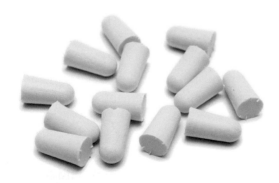

▲ Figure 1.51 Ear plugs

Noise at work can interfere with communications and make warnings harder to hear. It can also reduce a person's awareness of their surroundings. These factors can lead to safety risks and may put workers at risk of injury or death.

Risk assessment

Employers have a responsibility under the terms of HASAWA to carry out regular **risk assessments** to ensure there are minimal dangers to their employees. All workers should be properly trained in the precautions they need to take to avoid health and safety risks.

A **hazard** is anything that can cause harm. A **risk** is the chance, high or low, that somebody will be harmed by the hazard. Before using equipment and materials required for a job, it is necessary to first undertake the simple, but necessary step of a risk assessment (see Figure 1.52). Within a risk assessment you need to consider the planned work and think about what could cause harm to people. You can then decide whether you have taken enough precautions, or whether more should be done to prevent harm.

There are five steps to risk assessment:
1 Identify the hazards.
2 Decide who might be harmed and how.
3 Evaluate the risks and decide on whether the existing precautions are adequate or whether more should be done.
4 Record your findings and implement them.
5 Review your assessment and update if necessary.

Risk Assessment

Activity / Workplace assessed: Return to work after accident
Persons consulted / involved in risk assessment
Date:
Reviewed on:

Location:
Risk assessment reference number:
Review date:
Review by:

Significant hazard	People at risk and what is the risk Describe the harm that is likely to result from the hazard (e.g. cut, broken leg, chemical burn) and who could be harmed (e.g. employees, contractors, visitors)	Existing control measure What is currently in place to control the risk?	Risk rating Use matrix identified in guidance note Likelihood (L) Severity (S) Multiply (L) x (S) to produce risk rating (RR)				Further action required What is required to bring the risk down to an acceptable level? Use hierarchy of control described in guidance note when considering the controls needed	Actioned to: Who will complete the action?	Due date: When will the action be completed by?	Completion date: Initial and date once the action has been completed
			L	S	RR	L/M/H				
Uneven floors	Operatives	Verbal warning and supervision	2	1	2	M	None applicable	Site supervisor	Active now	Ongoing
Steps	Operatives	Verbal warning	2	1	2	M	None applicable	Site supervisor	Active now	Ongoing
Staircases	Operatives	Verbal warning	2	2	4	M	None applicable	Site supervisor	Active now	Ongoing

		Likelihood			Likelihood 3 – Very likely 2 – Possible 1 – Unlikely
		1 Unlikely	2 Possible	3 Very likely	
Severity	1 Slight/minor injuries/minor damage	1	2	3	Severity 3 – Major injury/extensive damage 2 – Medium injury/significant damage 1 – Slight/minor damage
	2 Medium injuries/significant damage	2	4	6	
	3 Major injury/extensive damage	3	6	9	

1 – Low risk, action should be taken to reduce the risk if reasonably practicable
2, 3, 4 – Medium risk, is a significant risk and would require an appropriate level of resource
6 & 9 – High risk, may require considerable resource to mitigate. Control should focus on elimination of risk, if not possible control should be obtained by following the hierarchy of control

▲ Figure 1.52 An example of a risk assessment

Method statement

Method Statement – Internal Decoration – Guidance notes	
The original template is available through HSEdocs.co.uk	
Scope of Works	

This method statement describes the work process for the following and must be completed before the start of an assignment.

1) **Start of works.** (This element may relate to health and safety aspects, protection of areas, selection of tools and equipment, etc.)

2) **Preparation of surfaces** (This element may be to remove existing wall coverings. List the method for removing wall coverings, preparing walls and applying size or similar sealer for next stage.)

3) **Hanging lining paper** (Step-by-step process)

4) **Hanging set patterned blown vinyl wallpaper** (Step-by-step process)

5) **Application of water-based paints to walls and woodwork** (Step-by-step process)

6) **Setting out borders** (Step-by-step process)

7) **Application of rag-roll effect** (Step-by-step process)

8) **Finishing off** (Describes how you want the site left at the end of the task, clearing away, reinstating furniture, etc.)

▲ Figure 1.53 Method statement template as used in the City and Guilds synoptic assignments

Method statements are widely used in the construction industry to help ensure safety and to communicate what is required to all those involved. A method statement is a useful way of recording the hazards involved in specific tasks and communicating the risk and precautions required to all those involved in the work. The statement need be no longer than necessary to achieve these objectives effectively. The method statement should be clear and can be illustrated by simple sketches. There is more information on method statements in Chapter 2.

Permit to work

When proposed work is identified as being high risk, strict controls are required. The work must be carried out against previously agreed safety procedures laid down in a permit to work. The permit to work is a documented procedure that authorises certain people to carry out specific work within a specified time frame. It sets out the precautions required to complete the work safely, based on a risk assessment. The permit to work document describes what work will be done and how it will be done, as detailed in the relevant method statement. A permit to work enables adequate control of the risks involved in a particular task.

The permit to work document requires signatures from those responsible for authorising the work and those carrying out the work. Where necessary, it also requires a declaration from those involved in shift handover procedures or extensions to the work. Finally, before equipment or machinery is put back into use a declaration from the permit originator that it is ready for normal use is required.

③ UNDERSTAND HOW TO SET UP AND SECURE WORK AREAS

Planning a site layout

▲ Figure 1.54 Construction site signage

Site layout plans are prepared by contractors as part of their setting-up activities before work begins on a construction site. They will be prepared by the CDM co-ordinator to ensure the safe and efficient running of the site. A site layout plan is a crucial part of construction management, as construction sites often involve the co-ordination and movement of large quantities of materials, people and high-value products and plant.

Site layout plans might show the location and size of areas for:
- zones for particular activities, such as assembly of elements of the building prior to erection and areas for welding steel. These activities should be allocated zones to allow them to work freely without disruption from other site activities

- cranes (including radii and capacities)
- site offices
- welfare facilities
- offloading, temporary storage and storage areas (laydown area)
- subcontractor facilities
- car parking
- emergency routes and muster points
- access, entrances, security and access controls, temporary roads and pedestrian routes
- vehicle wheel washing facilities
- waste management and recycling areas
- site hoardings and existing boundaries
- protection for trees, existing buildings and neighbouring buildings
- signage.

Site layout planning involves four basic processes:
- identifying the site facilities that will be required
- determining the sizes and other constraints of those facilities
- establishing the interrelationships between the facilities
- optimising the layout of the facilities on the site.

Figure 1.55 shows an example of a site layout plan.

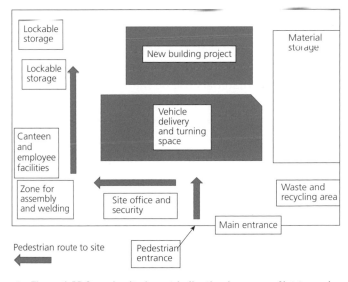

▲ Figure 1.55 Sample site layout indicating key areas. Not to scale

As sites will change in nature during the course of the works, there may be a number of different site layout plans for different phases. There may also be more detailed plans showing particularly complex areas or sequences or describing specific functions.

When planning the layout of the site, neighbouring properties, waste management activities, temporary services and environmental protection must all be considered.

Neighbouring properties

The plan should try to ensure that the construction site causes as little disruption to neighbouring properties as possible. Any disputes with neighbouring properties may cause delays and additional costs to a project. It is important to communicate the aims of the project and listen to any concerns neighbours may have, and where possible take steps to alleviate any potential upset.

Restricting noise to acceptable levels and carrying out work within reasonable hours of the day will demonstrate consideration for others. Entering and leaving the site may cause debris and mud to be carried off site onto the adjoining roads, so roads should be swept of debris. This is a local authority requirement as debris carried out on the wheels of various vehicles leaving the site could cause hazards on public roads. Considerate parking by contractors around the site is important, particularly if there is not room for a parking compound within the site boundary.

Waste management

Waste management activities are set out in a site waste management plan (SWMP) that is prepared by the client before construction begins. A SWMP describes how materials will be managed efficiently and disposed of legally during the construction work and explains how the reuse and recycling of materials will be maximised. During the construction phase of a project the main contractor will need to ensure that the waste management plan is enforced and must supply different waste containers for segregating different materials. Many of these, such as plasterboard, brick rubble and timber, can be recycled by specialist recycling agents. Hazardous waste should be kept separate and arrangements made with contractors for regular removal of the various waste products.

Environmental protection

Protecting the environment is essential, and this may require trees, local flora and fauna to have coverings or barriers placed around them. Consideration and protection should be put in place to ensure that any wildlife that may be disturbed from its habitat is protected or relocated.

Temporary services

Water, electricity, drainage and telephones will be required on site during the construction process. Money can be saved if these can eventually form part of the services for the proposed buildings.

Routes around site

Roads may be needed to move materials around the site. There are three ways of providing roads during construction:

- Permanent roads can be laid before the construction process starts. This means completing the installation of all services before roads are laid. However, there is a risk of damage to roads by heavy plant or traffic during the construction process.
- A hardcore base can be laid along the route of the permanent road. Traffic and water may damage the subsoil below the hardcore, and this could mean having to replace the hardcore and the damaged soil.
- Temporary roads can be laid along the most convenient routes to suit construction. This will reduce some of the expenditure related to the permanent road features, as the base will already be in place.

▲ Figure 1.56 Barriers in place to provide clear routes around a construction site

Site welfare

Areas of welfare need to be considered during site set-up, and will include the following facilities:

- toilets
- washing facilities
- storage for personal items
- canteen
- drying room.

▲ Figure 1.57 Welfare facilities

Site safety and security

It is important to ensure that work areas are protected in relation to the safety and security of the public and employees. A fence or hoarding should be erected around the site to prevent the public from wandering onto the site and to protect them from building work. Permission must be obtained from the local authority highways department and the police for any temporary pavement crossing. The cost of reinstating the pavement must be paid to the local authority.

Protecting the public

In the past, construction sites have attracted children who may view the site as a playground. For safety reasons, the public, including children, must be kept out of the construction site area. Fencing, hoardings, signage and barriers, as well as security personnel at the site entrance, can all be used to prevent unauthorised access and protect the public.

▲ Figure 1.58 Barriers used to keep pedestrians away from work and an open trench

▲ Figure 1.59 Metal fencing, sometimes called Heras after the manufacturer, is being used around this site to keep unauthorised people out

Protecting employees

Employees should also be protected from harm as there will be hazards on a large construction site. Cranes, lorries and other plant equipment moving around the site need sufficient barriers and procedures in place to prevent accidents. Open extractions (such as drainage channels) and various trip hazards may also be found on a building site and every precaution should be taken to avoid harm to employees.

▲ Figure 1.60 Pedestrian route around construction works

Pedestrianised areas provide construction personnel with a safe passage to allow them to move around the site while keeping clear of moving vehicles. Informing personnel when various types of equipment are being operated on a day-to-day basis as well as carrying out toolbox talks will help to keep everyone informed of various safety hazards. A site induction will be carried out for every person working on or visiting the site to inform them of the rules and processes operating on that particular site.

IMPROVE YOUR ENGLISH

In pairs, discuss and write down ten items that should be included in a site induction and five topics that could be discussed during a toolbox talk. The HSE website has some templates for both inductions and toolbox talks for construction: www.hse.gov.uk/construction/resources/toolboxtalks.htm

Use the HSE website to check your answers.

Site security

Consideration must be given to the types of material to be stored on the construction site, and what will be required to protect them from damage, misuse, theft and the effects of weather. Where possible, storage areas should be sited to prevent double handling, for example bricks could be offloaded and stored at different points around the site.

▲ Figure 1.61 Demountable buildings

All deliveries should be checked in when they arrive on site and compared against orders to identify any omissions or discrepancies. All delivery notes should be filed to allow for checking cost against the invoices for the materials. Any discrepancies in deliveries should be reported to the supplier so they are quickly informed and replacements may be sent out. A record of the call should be kept in the site diary, which should be kept up to date with all incidents/events that occur on site.

4 KNOW BUILDING SUBSTRUCTURE

The Building Cost Information Service (BCIS) defines the term 'substructure' as: 'All work below underside of **screed** or, where no screed exists, to underside of lowest floor finishes including damp proof membrane, together with relevant excavations and foundations (includes walls to basements designed as retaining walls).' The BCIS states that the function of the substructure is to: '... transfer the load of the building to the ground and to isolate it horizontally from the ground'.

KEY TERM

Screed: a levelled layer of material (e.g. cement) applied to a floor or other surface.

Purposes and materials of substructure

This section focuses on the purposes of and materials used in the substructure of a building.

Types and purposes of foundations

Foundations provide support to buildings and other structures and transfer the load of the building to solid ground that is capable of supporting the weight. Very broadly, foundations can be categorised as shallow foundations or deep foundations. The main types of foundation considered in this section are strip, pile, pad and raft, as explained in the table opposite. The type of foundation is selected after considering a number of factors, including:

- the nature of the load to be supported
- ground conditions
- the presence of water
- space availability
- accessibility
- sensitivity to noise and vibration.

Strip foundations

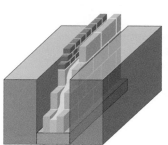

Traditional strip foundation

Trench foundation with reinforcing bars for extra strength

- Strip foundations provide a continuous strip of support around the perimeter of the building
- In some cases, foundations will also be taken within the building area to provide support for internal load-bearing walls
- They are sometimes referred to as trench foundations, as the whole trench is filled to the top with concrete

Pad foundations

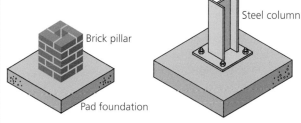

- Pad foundations are rectangular or circular pads that are used to support localised loads from columns
- They are typically used in a series, e.g. in a frame structure where the bearing capacity of the ground is sufficient at low depth
- They are not suitable for heavy structures that are prone to being affected by wind forces or excessive ground movement

Pile foundations

- Pile foundations are deep foundations. They are formed of long, slender, columns, typically made from steel or reinforced concrete
- Piles are used when there is a need to go to deep levels to find solid ground for the foundation. They will typically be used in buildings sited near rivers, as the piles will be required to find bed-rock below the level of the water table. Large buildings, such as office buildings, will also be built on top of pile foundations
- The most common types of piles are driven piles that are prefabricated off site and then driven into the ground, and bored piles that are poured in situ. The piles will be set out in series at the points where the main loads of the building will need to be transferred

Raft foundations

- Raft foundations are formed by pouring concrete or sometimes reinforced concrete slabs of uniform thickness (typically 150 mm to 300 mm) that cover a wide area, usually covering the entire area of the building. The entire load of the building is then transferred over the wide area much as in the way a raft floats on water

Characteristics of materials used in substructure

Brick 	• Bricks are small rectangular blocks that can be used to form parts of buildings, typically walls • The use of bricks dates back to pre-Roman times, when they were made from clay dug from the ground. Handmade bricks are still required when carrying out work on historic buildings. Since the time of the Industrial Revolution, bricks have generally been machine-made with clay taken from quarries in various parts of the United Kingdom. The clay varies in colour and strength and the designer needs to consider these factors when choosing bricks • Bricks are in common use for the construction of walls, paving and more complex features such as columns, arches, fireplaces and chimneys. They remain popular because they are relatively small and easy to handle, can be extremely strong in compression, are durable and low maintenance, can be built up into complex shapes and are visually attractive
Block 	• Blocks are denser and larger than bricks. Solid concrete blocks are manufactured from naturally dense aggregates to be strong and heavy. Solid concrete blocks are strong enough to be used for large masonry units that are load bearing • Block density depends on what they are to be used for; they can be reasonably lightweight when used for interior walls that are not load bearing • As blocks are larger, they can be used to build a large area quite quickly, so if they are not required to be decorative then they are selected instead of bricks due to their speed of laying
Steel 	• Steel is used to provide **reinforcement** to the substructure in the form of mesh or reinforcing bars • The main purpose of steel is to strengthen and hold concrete in tension and reduce cracking in the concrete
Concrete 	• Concrete is the most used man-made material in construction. It is used extensively in buildings, bridges, tunnels, roads and dams. Concrete is mainly used for foundations, floors, paths, kerbs and other items such as large underground sewers • Some items may be prefabricated off site and, depending on their use, they may have steel reinforcement added • Concrete is a composite material, consisting mainly of Portland cement, water and aggregate (gravel, sand or rock). When these materials are mixed together, they form a workable paste which gradually hardens over time

Damp proof course/damp proof membrane

Damp proof course (DPC)

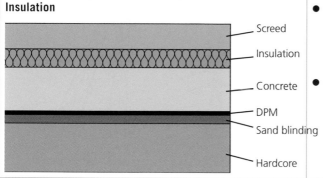

Damp proof membrane (DPM)

- A damp proof course (DPC) is laid between courses of bricks to prevent damp rising vertically through the bricks into the building. Rising damp is caused by **capillary action** drawing moisture up through the porous elements of a building's fabric
- A DPC is supplied on a roll and Building Regulations state that a damp proof course may be a bituminous material, polyethylene, engineering bricks or slates in cement mortar or any other material that will prevent the passage of moisture
- A damp proof membrane (DPM) is a membrane material applied to prevent moisture transmission. Typically, a DPM is a polyethylene sheet laid under a concrete slab to prevent the concrete from taking up moisture through capillary action. The Building Regulations state that a ground-supported floor will meet these requirements if the ground is covered with dense concrete laid on a hardcore bed and a DPM is provided

Insulation

- Screed
- Insulation
- Concrete
- DPM
- Sand blinding
- Hardcore

- In the average home, approximately 10 per cent of heat loss is through the ground floor. Insulation is a means of reducing heat loss and is typically installed on the ground floor between the top screed and concrete layer. Insulation may also be used in upper floors between heated and unheated spaces, such as a loft
- Building Regulations require that floors achieve minimum thermal performance, and this affects the amount of insulation required. Thermal performance is measured in terms of **U-values**

KEY TERMS

Reinforcement: the action or process of strengthening.

Capillary action: the ability of water to rise up microscopic tubes in building materials (e.g. most brick types, some stones, concrete blocks and plaster). When these materials are in contact with moisture, the water adheres to the pores of the material's capillaries or microscopic tubes and rises up to cause damp.

U-values: a measure of how quickly heat will travel through the floor leading to loss of heat. Different materials have higher or lower U-values and therefore their selection will depend on what they are being used for.

Types of building services

The majority of buildings being constructed will need to be connected to services such as electricity, gas, water and drainage, and communications before they can be used.

Electricity

Electricity is supplied to buildings via the National Grid, which is a network of pylons, cables, sub-stations and transformers that are used to transfer power from the source that is generating electrical energy. Sources of electric energy include coal, gas and nuclear power stations and renewable sources such as solar panels, wind turbines, hydro or tidal power, and also electric storage batteries. A mains cable is laid into the buildings, usually through underground ducts, to a meter inside the building.

The electrician then runs cables from this single source to various points around the building to provide power for appliances, lighting and other items for the householder or building user.

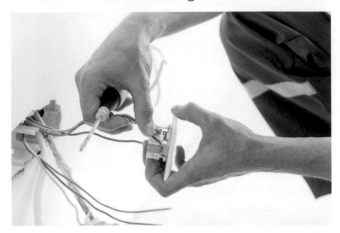

▲ Figure 1.62 Electrical socket

▲ Figure 1.63 Electricity pylons

Gas

Natural gas is supplied from the gas refinery to homes and other buildings via gas mains, which are the pipeline infrastructure. In the UK, mains gas is supplied to more than 21 million homes and it is the most popular fuel for heating and cooking. The gas arrives at the individual's home through underground pipes to a metered supply. The heating engineer will run pipes from this single-entry point to the various parts of the building, wherever there is a requirement for gas.

▲ Figure 1.64 Gas valves

▲ Figure 1.65 Gas pipes being connected to a gas boiler.

Water

The water coming out of your tap started its journey falling as rain, hail or snow. Water is collected in reservoirs and in underground aquifers before being transported via pipes to the treatment works. Water mains pipes carry the water to the roads next to a building. From here a water supply pipe connects the water main to each property. The water supply will feed the heating system, kitchen and bathroom taps and toilets as well as appliances like washing machines and dishwashers.

▲ Figure 1.66 Water reservoir

▲ Figure 1.67 Plumbing equipment

Drainage (surface and wastewater)

Drainage is required in buildings to take away the water and soil waste. Drains are underground pipes that take water away from houses and buildings. Most properties have separate drains for rainwater and wastewater. Properties are usually connected to mains drainage which is a network of pipes that collect wastewater and then take it via underground sewers to the sewage treatment works.

Surface water drainage

Surface water drainage refers to the drains that collect rainwater from roofs, driveways and roads, also known as 'storm water'. Rainwater is untreated and is usually taken by surface water drains directly to rivers and outlets near the sea. Rainwater may also be collected by pipes and directed to the main drains to travel to the water and sewage treatment plant before its journey back into the water supply system or to be displaced in rivers and the sea.

▲ Figure 1.68 Drainage system

RAINWATER HARVESTING

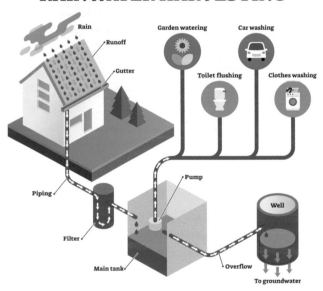

▲ Figure 1.69 Rainwater harvesting system

A more environmentally friendly method of collecting rainwater is to use a rainwater harvesting (recycling) system (see Figure 1.69), where the water is filtered before being stored for use in the garden, for car washing and even to flush toilets.

Wastewater drainage and disposal

▲ Figure 1.70 Soil water drainage

Wastewater or soil water comes from toilets, sinks, baths and showers, washing machines and dishwashers, and the pipes from these will discharge into a drain. Wastewater may also be referred to as 'foul water'.

The soil water drain takes the foul water via the main sewers to the local wastewater treatment works. Sometimes rural areas are not connected to mains drainage and rely on their waste being treated within septic tanks before the treated water is released back into the land. The solid waste that forms in the tanks is periodically removed, either for use as fertiliser or transferred to the main treatment works.

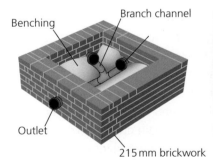

▲ Figure 1.71 Benched drain

▲ Figure 1.72 Septic tank

▲ Figure 1.73 Sewage treatment plant

A 'grey water' recycling system involves collecting bath and shower water which is then filtered, treated and recycled so it can be used for toilet flushing and to water the garden. This is a way of reusing some of the soil wastewater but is not yet a requirement for new buildings. Grey water recycling systems are considered to be more environmentally friendly as they reduce water usage.

Communication networks

▲ Figure 1.74 Underground cables

▲ Figure 1.75 Installing aerial cables

Communication networks supply the internet, television, telephones and other means of data or speech transfer. The majority of this is brought to buildings via underground cables that pass through ducts to separate them from one another. Signals are generally transmitted from radio, television and internet providers via satellites or aerials and other transmitting masts that are erected on high ground or buildings.

IMPROVE YOUR ENGLISH

1 List three methods of renewable energy sources that produce electricity.
2 What is the benefit of a rainwater harvesting system?

⑤ KNOW BUILDING SUPERSTRUCTURE

The superstructure is the part of a building above its foundations. A superstructure includes the following:

- **Frame**: load-bearing framework, main floor and roof beams, ties and roof trusses of framed buildings, casing to stanchions and beams for structural or protective purposes.
- **Upper floors**: suspended floors over or in basements, service floors, balconies, sloping floors, walkways and top landings where part of the floor rather than part of the staircase.
- **Roof**: roof structure, roof coverings, roof drainage, rooflights and roof features.
- **Stairs and ramps**: construction of ramps, stairs, ladders, etc., connecting floors at different levels.
- **External walls**: external enclosing walls including walls to basements, but excluding walls to basements designed as retaining walls.
- **Openings in external walls**: windows and doors.
- **Internal walls**: includes internal walls, partitions, balustrades, moveable room dividers, cubicles and the like.
- **Openings in internal walls and partitions**: doors, hatches and other openings.
- **Internal services and fixtures and fittings**: may be included as part of the superstructure as they are fitted above ground.

Wall types

Solid walls

Solid walls may be made from brick, block, stone or concrete and are a traditional form of wall construction that was used regularly for the main structure of a house in new constructions up until the early 1900s.

Some of the advantages of solid walls are they provide good sound insulation and the building remains cooler in the summer. The main disadvantage of using solid walls in modern buildings is that they are prone to allowing damp to permeate, either through rising damp or being absorbed through the walls.

Solid walls are still used in construction, particularly for garden wall construction.

▲ Figure 1.76 Solid wall

Cavity walls

Cavity wall construction was introduced in the UK around 1920 and resulted in significant improvements to buildings, particularly their ability to exclude damp.

Cavity walls also provide space for insulation to reduce heat loss through walls.

Advantages of cavity walls are:

- they reduce the weight on the foundation
- they provide good sound insulation and help to keep out noise
- they give better thermal insulation than a solid wall because the space is full of air or insulation materials and this reduces heat transmission (they have a heat flow rate that is 50 per cent less than that of a solid wall)
- economically cheaper to build than solid walls
- they are fire resistant.

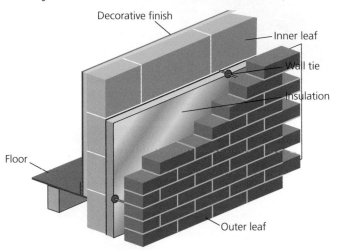

▲ Figure 1.77 Cavity wall

Timber frame

The term 'timber frame' typically describes a system of panels constructed from small-section timber studs to form structural walls and floors. The frames are clad with board products and filled with insulation material. Often brick or other cladding material is built on the outer face. Timber frame construction is an efficient and quick method and is favoured by modern housebuilders. Houses can be quickly erected on site and made watertight so that the inside of the building can be worked on sooner. Most timber frames are manufactured off site in factories, then delivered to site and hoisted into position by cranes. Once all panels are erected, the upper floors and roof can be added. The exterior cladding is then added and the interior can be completed.

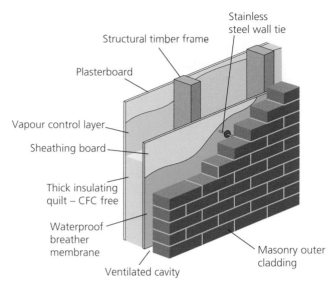

▲ Figure 1.78 Timber framing

▲ Figure 1.79 Timber frame panels lowered into position by crane

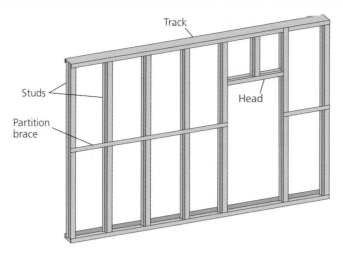

▲ Figure 1.80 Partitions: metal stud wall

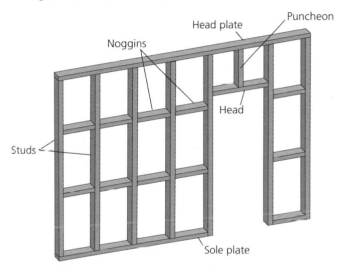

▲ Figure 1.81 Partitions: timber stud wall

▲ Figure 1.82 Thin metal stud walls faced with plasterboard

Partitions

Partitions are typically used to divide up the space inside a building into separate rooms. They can be made from brick, lightweight block, prefabricated panels, metal studwork and timberwork. Partitions do not carry any loads from upper floors or the structure of the building: this is referred to as 'non-load-bearing'.

Many modern house styles, particularly timber frame, may use prefabricated panels for interior partition work, but they can have timber studwork built on site and then faced with plasterboard.

Other types of materials used for superstructures

Stone 	• Stone has been used for the superstructure of buildings for centuries as it was often found close to proposed construction sites. More commonly used today for decorative panels and garden walls • It is more time consuming to build with stone than brick, timber or block, so is a more expensive process
Timber 	• Timber has been used extensively over the centuries to form the superstructure of a building. The most notable timber houses were built in the Tudor and Elizabethan periods • Modern housing methods often use timber frames and cladding as an alternative to brick buildings
Metal stud 	• Metal stud or studding refers to the construction of thin walls where the frame construction is made from thin strips of metal which are then fixed together to form a framework • The illustration shows thin metal stud walls forming an interior space ready to be faced with plasterboard or another covering material

Floor types

Floors are used at various levels in a building, depending on the height and use of the building. Most common domestic housing is built over two or three levels, to provide a ground, first and possible second floor level. Taller buildings such as flats and offices may have multiple floor levels.

Solid floor

Figure 1.83 shows a section through a typical ground floor construction and indicates the various layers used to make a floor that can withstand moisture penetration and reduce heat loss through the floor. Layers shown are a **hardcore** solid bed of stone chippings compressed to a level surface and covered with a layer of sand referred to as **blinding**. Blinding fills the voids that may be left between the chippings. The **damp proof membrane (DPM)** is laid over this levelled sand layer to prevent rising damp and then **concrete** is poured and levelled over the top. Sometimes **steel reinforcement mesh** may be laid within the concrete if a floor needs to be able to withstand greater loadings than a normal domestic house. An **insulation** layer (typically a rigid foam board) is laid over the dry concrete before the **screed** layer is applied. The screed layer provides a final levelled surface that is then suitable for the floor covering such as carpet or tiles to be laid.

Steel reinforcement (also known as **rebar**, which is short for reinforcing bar) is steel bars or a mesh of steel wires used in reinforced concrete and masonry structures to strengthen and hold the concrete in tension. To improve the quality of the bond with the concrete, the surface of rebar is often patterned.

▲ Figure 1.84 Rebar

Suspended floor

A suspended floor is an alternative method of providing a ground floor construction and can be seen in the section drawing in Figure 1.85. This is shown with a cavity wall. The internal floor level should be fixed above the external ground level. Where timber joists have been used to support the floorboards there should be a through flow of air provided under the floor by an air vent in the wall, as this will reduce the likelihood of the timbers becoming damp and decaying.

A joist is a horizontal structure that is used to span an open space, in this illustration from the honeycomb sleeper wall to the joist hanging brackets that are fixed in the wall.

Floorboards or sheet timber are used to level the floor, so that carpets and tiles may be laid.

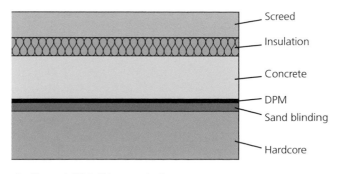

▲ Figure 1.83 Solid concrete floor

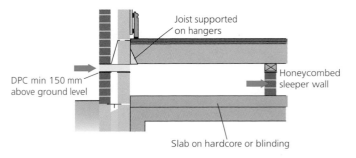

▲ Figure 1.85 Suspended wood floor

Honeycomb sleeper wall

A honeycomb sleeper wall is a brick wall with openings, created either by allowing gaps between stretchers or by omitting bricks. It is used to support floor joists and provide ventilation under floors.

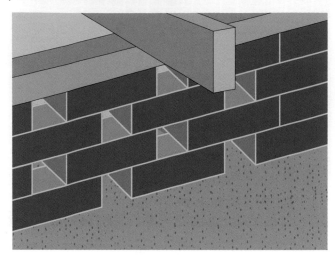

▲ Figure 1.86 Honeycomb sleeper wall

Block and beam constructed floor

A block and beam constructed floor uses **precast beams** and blocks that are predesigned for the length of the rooms or spaces, supported on concrete blockwork to provide a solid underfloor (see Figure 1.87). The upper surface of this floor then has insulation boards laid before being screeded (as described earlier, in the section on solid floor construction).

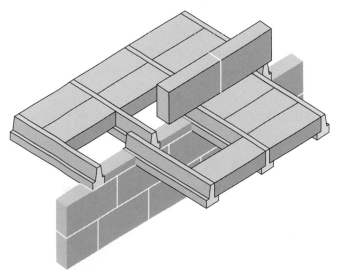

▲ Figure 1.87 Block and beam floor

Precast floor

A precast floor is a method in which concrete floor panels are manufactured in the factory to a predesigned size. Again, there will be steel reinforcement material laid within the floor panels during manufacturing. Precast flooring is often used in offices and tall buildings, especially where it is difficult to pour concrete to form the floor.

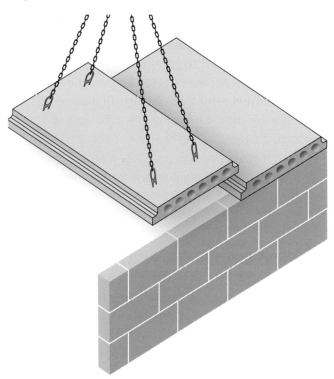

▲ Figure 1.88 Precast floor

Roof types

This section provides an overview of different types of roofs and their components.

Pitched roof  Double pitch roof  Lean-to roof	• A pitched roof is a roof that slopes downwards, typically in two parts, at an angle from a central ridge board • A pitched roof can slope on one side only, as in a lean-to roof or mono pitch roof
Traditional hand cut roof 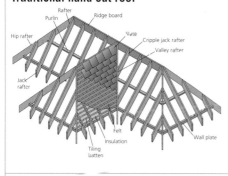	• A traditional hand cut roof is a roof that has been cut and assembled on site, rather than manufactured off site as in a trussed roof • The illustration shows the different parts of a traditional cut roof called a hipped roof. In a hipped roof (or hip-roof), all sides slope downwards to the walls
Trussed roof 	• A trussed roof is typically manufactured off site in a factory and then delivered and erected on site with the help of cranes to lift it into position • The roof is made up of a series of pre-built frames called trusses that are assembled to form the roof structure, before battening, felting and the application of the final roof covering
Flat roof 	• A flat roof is a roof that is almost level. To be classed as a flat roof, the pitch or slope of the roof must be no more than 10 degrees to the horizontal This angle no more than 10°

Roof materials

Timber 	• Timber is usually used in a structural form to provide the framework to support the roof coverings • The illustration shows the timber rafters, support battens and timber sheet material that is visible before the waterproof tiles are fixed in place
Lead 	• It is possible to completely cover a roof in lead and many churches have this type of roof. However, for most domestic and commercial buildings, it is commonly used for smaller areas such as flashings and lead gulleys • The illustration shows lead flashing being used to weatherproof the joint between the sloping roof and the wall
Slate 	• Slate has been used as a roof covering for hundreds of years. It is a natural rock that is mined and cut to form thin sheets that are shaped to size and then overlapped to keep the weather out • Slate and stone roofs are a feature of many historic buildings. To repair and conserve them successfully requires the use of appropriate traditional materials and techniques
Tile 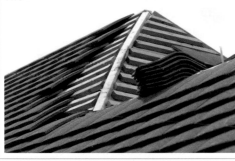	• Clay tiles have been used to tile roofs for hundreds of years. Originally handmade, the tiles were traditionally made from clay in moulds, although concrete has been used since the 1950s • Tiles need to be laid in an overlapping pattern to be completely weatherproof
Felt 	• Felt can be used in roofing in two ways: as an outer waterproof covering, or as an underlayer to other roofing materials such as slates and tiles • Felt helps to reduce the effect of wind loading on the slate or tile roof covering. It also provides a waterproof barrier and allows for the safe disposal of water that has collected on the upper surface of the underlay • The illustration shows the black felt underlay fixed to the rafters under the battens and tiles
Sheet 	• The illustration shows heat-applied bituminous felt roofing materials. Typically, three layers are bonded to the roof substrate using heated bitumen • Sheet timber material is also used under the tiles to improve insulation and provide additional security

Other roofing material systems	• Metal, including zinc, powder-coated steel and plastic reinforced sheeting, is also used for covering roofs but these are more likely to be used on industrial buildings than domestic ones • Hard plastics are manufactured to be used in the place of clay and concrete tiles and are ideal for off-site manufacture of buildings as they are a much lighter material and lend themselves well to prefabrication
 Metal sheet roofing	
 Rubber roofing sheets	• In many modern buildings, synthetic systems of covering are used to replace, in particular, bituminous felt on flat roofs. The most popular being rubber sheeting (ethylene propylene diene monomer (EPDM)) or glass reinforced plastic (GRP). Rubber roofing comes in pre-made rolls, ready to be rolled out across the roofing. A relatively simple installation process involves a single sheet of rubber-like material, which is glued to the roof decking, and then fitted with trims around the edges • To install glass reinforced plastic roofing, a layer of resin is added to the roof surface, then a mat of fibreglass strands, and then another layer of resin. A topcoat is then applied afterwards to create a weatherproof surface

Other roofing material systems

Metal, including zinc, powder-coated steel and plastic reinforced sheeting, is also used for covering roofs but is more likely to be used on industrial buildings than domestic ones.

Hard plastics are manufactured for use instead of clay and concrete tiles. They are ideal for off-site manufacture of buildings as they are a much lighter material and lend themselves well to prefabrication.

In many modern buildings, synthetic systems of covering are used to replace bituminous felt on flat roofs. The most popular are rubber sheeting called EPDM (ethylene propylene diene monomer) or glass-reinforced plastic (GRP).

Rubber roofing comes in pre-made rolls, ready to be rolled out across the roof. It is a relatively simple installation process and involves a single sheet of rubber-like material that is glued to the roof decking and then fitted with trims around the edges.

To install glass reinforced plastic roofing, a layer of resin is added to the roof surface followed by a mat of fibreglass strands and then another layer of resin. A topcoat is then applied to create a weatherproof surface.

Types of finishes

This section provides an overview of the types of finishes encountered in a typical building. More detail is provided on paint systems and paper coverings in later chapters.

Internal finishes

Paint systems

Paint systems are described in detail in Chapters 3 and 4, but in simple terms paint systems are selected to meet the following criteria:

- **Preservation**: to provide protection and make the surface to which they are applied last longer.
- **Sanitation**: to make the surface easy to clean by using coatings that can be washed down.
- **Decoration**: to provide different colours to appeal to individual tastes, which can be changed when required. (Further detail on the use of colour in decoration can be found in Chapter 5.)
- **Identification**: paints can be used to make items stand out from one another and in particular to identify parts, for example British Standard 1710 provides the standards for identifying pipes (there is more detail on this in Chapter 4).

Paper coverings

Paper coverings, such as wallpapers, are used to provide, colour, texture and pattern to interiors, although some papers provide a preparatory role, for example lining papers (see also Chapter 5).

Plaster

Plaster is used on interior surfaces to provide a sound surface on plasterboard and brick/blockwork walls for paint to be applied. Plaster is relatively slow drying and produces a lot of moisture while drying out.

Dry lining

Dry lining is a system applied to the internal faces of buildings, such as walls and ceilings. Plasterboard is attached to the substrate to form a smooth surface that finishes such as paint can be applied to directly; a 'wet' plaster finish is not required. The joints of the boards are normally taped, filled and sanded before coating with paint.

Tiling

Tiling involves fixing ceramic tiles to a surface. Tiles are typically fixed to the walls of shower cubicles and around baths, basins and kitchen worktops to provide a waterproof finish which can be easily wiped down. The tiles are fixed with a waterproof adhesive and the joints in the tiles are filled with waterproof grout to ensure that no water penetrates to the substrate.

▲ Figure 1.89 Large glazed tiles being fixed to the interior

External finishes

The main purpose of an external finish is to provide protection to the building as a whole. Some examples of systems that are commonly used are described below.

Paint systems

The primary role of an external paint system is to preserve and protect against weathering. External paint also provides a decorative colour and is more sanitary as it can be washed when required. Paint systems for exterior surfaces are discussed in more detail in Chapters 3 and 4.

Rendering systems

Traditional render systems have been around for hundreds of years. An example of a traditional render system is sand and cement render. Materials for a sand and cement render are cheap and the system can be repaired easily when cracks appear. The system is effective at protecting walls as well as covering poor and damaged brickwork.

Modern render systems are self-coloured (meaning they do not require painting) and have many benefits. Silicone render is an example of a modern render system, because the inclusion of silicone makes the render more flexible.

Special coatings

Anti-graffiti paints, heavy epoxy-based wall coatings, antibacterial coatings, thermal paints, micaceous oxide, water repellents, anti-burglar paints and flame retardants are some of the specialist coatings designed for a specific purpose. These coatings do not follow the same preparation and application processes of ordinary paint coatings.

External wall insulation

Insulation can be used to reduce the transmission of heat from the inside and outside of an enclosed space such as a building. External wall insulation (EWI) is the application of thermal insulation to the external walls of buildings and a finish system to the outside face of the external walls of an existing building to improve its thermal performance.

Cladding

The term 'cladding' refers to components that are attached to the primary structure of a building to form non-structural, external surfaces. Different types of materials can be used for this process, such as timber, metal, concrete, brick and tile.

▲ Figure 1.90 Timber cladding being fixed as an outer finish

Building elements

The construction of a building is made up of a series of elements and these are categorised as first fix or second fix.

First fix

First fix relates to the initial or first fixing of items within a building that need to be carried out before other building processes can take place. Generally, first fix will include:

- building the structure and outside cladding, including roof tiles, to make the building watertight
- roof joists
- floor joists and flooring
- door frames and door linings
- partitions and staircases
- electrical cabling and pipework for plumbing and gas.

Typically, timber, metal and plastics are used to construct these building elements, the majority of which will be unseen when the building has been completed.

Partitions

As mentioned earlier in this chapter, partitions are used to divide the internal spaces of the building and may be constructed from timber, metal framing, brickwork or blockwork.

External door and window frames

These are the outer frames before doors or glass are fitted. However, it is more common for the glazing units for windows to be fitted before moving on to the second fix in order to make the building more secure and also watertight.

Internal door lining

An internal door lining is the frame in which a door will be fitted. The lining is fixed to the internal walls or partitions to form door openings. The example shown in Figure 1.91 has the door already fitted in the opening and **ironmongery** fitted.

KEY TERM

Ironmongery: items such as door handles, locks, latches and hinges that are usually made from iron, steel, aluminium, brass and plastics.

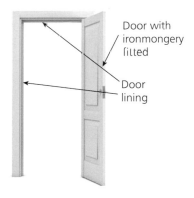

Door with ironmongery fitted

Door lining

▲ Figure 1.91 Door with ironmongery

Stairs and staircases

These allow the occupants of the building to access other floor levels within the building and are usually constructed from timber, metal, glass, stone or concrete. Figure 1.92 on the next page shows the key parts of a staircase.

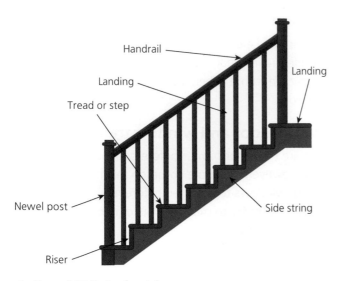

▲ Figure 1.92 Parts of a staircase

Services

Services in buildings refers to the supply of electricity, gas, communication networks (telephone, internet and TV), drainage and water supply. Although roads and other structures such as drainage may be referred to as services, they are more usually referred to as **infrastructure**.

KEY TERM

Infrastructure: the basic systems and services, such as transport and power supplies, needed for people to live and work effectively. Mains power, gas, water, drainage and communications are laid to a construction site to enable the various connections to be made by the electricians and plumbers, etc.

Second fix

Second fix is the term used to describe finishing processes within a building. Completion of plastering and dry lining will normally lead to the start of the second fix. Tasks such as the application of paints (painters), installing switches, sockets and lighting (electricians), fixing doors, ironmongery and kitchen units (carpenters), installing bathroom suites, sinks and heating (plumbers), applying wall and floor tiles (tilers) are all part of second fixing.

Finishes

This refers to processes that add a decorative and protective finish to the building and includes the application of paints, wall and floor tiles and decorative panelling.

Doors

Doors are fitted to provide security, access and egress to a building and to retain warmth. A variety of styles and construction methods are used, including timber, metal and plastics.

Kitchen units

Kitchen units are available as floor-standing base units and wall units, as well as units designed to fit in tall and narrow spaces. They can include an integral space to accommodate the cooker, refrigerator, dishwasher and washing machine. The kitchen sink may be inset into the worktop that is attached to the top of the units. Worktops provide a surface for the preparation of meals.

Sanitary ware

Sanitary ware refers to baths, showers, bidets, basins, sinks and toilets. Items are mostly constructed from pottery and ceramics but can also be made from hard plastics.

Figure 1.93 shows the location of first and second fixes. Red circles indicate first fix items such as roofing, flooring, partitions windows, stairs, etc. Blue circles indicate second fix items such as kitchen units, sanitary ware, lighting and doors, etc.

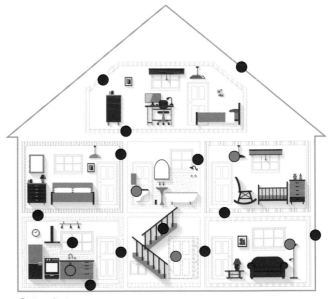

● First fix items
● Second fix items

▲ Figure 1.93 First and second fixes

Test your knowledge

1 Which description describes a quantity surveyor?

   a Concerned with project cost to provide estimates, budget and cost control.

   b Concerned with measured survey and design, structural surveys, legal matters.

   c Concerned with the strength and stability of the elements of construction.

   d Concerned with the plan and design and will oversee the construction of buildings.

2 What elements need to be completed during the first fix?

   a Ceiling roses, heating and water supply must be finished first.

   b Ceilings, walls and floors are finished first.

   c Stairs, services and light switches are finished first.

   d Finishes, stairs and services are finished first.

3 A paint delivery has missing items. How should this be reported to the supplier?

   a Text message   c Telephone

   b Letter   d Fax message

4 Which of the following is essential to be shown on a working drawing?

   a Colours   c Schedule

   b Quantities   d Scale

5 Which construction work is classed as industrial?

   a Painting houses   c Painting bridges

   b Painting offices   d Painting flats

6 Which of these members of a building team is referred to as the craftsperson?

   a Site agent   c Structural engineer

   b Electrician   d Chargehand

7 Which of the following walls is classed as load bearing?

   a Metal stud partition

   b Timber stud partition

   c Dense concrete block

   d Lightweight concrete block

8 What material would prevent moisture from seeping into the building through the floor?

   a DPM   c Screed

   b Block   d Concrete

9 What regulation states that injuries resulting in more than seven days' absence must be reported?

   a PUWER   c RIDDOR

   b HASAWA   d CDM

10 What hazard does the symbol below represent?

   a Flammable   c Explosive

   b Toxic   d Corrosive

Practice assignment

Working in pairs use information from the HSE website and this textbook to select a health and safety topic. Produce a short presentation that could be used as a toolbox talk. Include diagrams and if you have access to a computer, this could be produced as a PowerPoint presentation.

CHAPTER 2

ERECT AND USE ACCESS EQUIPMENT

INTRODUCTION

A large proportion of a decorator's work can be carried out from ground level and much can be done without steps or other access equipment, particularly when working on interiors. However, there will be many occasions when access to higher levels or areas such as the open well of a staircase is needed, so you will need a good understanding of the principles of working at height.

There are various types of equipment you can use for working at height. In this chapter we will consider the use of ladders, stepladders, leaning/standing ladders, platform steps, trestle platforms, podiums/hop-ups and mobile tower scaffolds.

By the end of this chapter, you will have an understanding of:
● selecting and inspecting access equipment
● using access equipment.

The table below shows how the main headings in this chapter cover the learning outcomes for each qualification specification.

Chapter section	Level 1 Diploma in Painting and Decorating (6707-13) Unit 116	Level 2 Diploma in Painting and Decorating (6707-22/23) Unit 220	Level 2 Technical Certificate in Painting and Decorating (7907-20) Unit 202	Level 3 Advanced Technical Diploma in Painting and Decorating (7907-30) Unit 302	Level 2 City & Guilds NVQ Diploma in Decorative Finishing and Industrial Painting Occupations (6572-20) Unit 224
1. Understand selection and inspection of access equipment	1.1–1.4, 2.1–2.4, 3.1–3.6, 4.1–4.4	1.1–1.3, 2.1–2.2, 3.1–3.4, 4.1–4.4	1.1–1.2	1.1–1.2	1.1–1.4, 2.1–2.3, 3.1–3.3, 4.1–4.5
2. Use access equipment	5.1–5.4, 6.1–6.2, 7.1–7.2	5.1–5.5, 6.1–6.4, 7.1–7.2, 8.1	2.1–2.2	2.1–2.2	5.1–5.5, 7.1–7.6

1 UNDERSTAND SELECTION AND INSPECTION OF ACCESS EQUIPMENT

Before selecting access equipment, you should fully understand how to work safely, including safe methods of working at height, and your responsibilities under the terms of the relevant regulations. This chapter provides information on some of this legislation and how it provides guidance when working at height.

Working at height

Access equipment enables you to gain access to work at a higher level than you can reach from the floor. Using access equipment and working platforms involves working at height, which means there is a risk of injury or even death as a result of a fall. By taking the correct **precautions**, this risk can be minimised.

▲ Figure 2.1 Fixed scaffolding on an office building

KEY TERM

Precautions: measures taken in advance to prevent something dangerous happening.

Health and Safety Executive

The Health and Safety Executive (HSE) helps businesses to keep people safe at work by providing guidance and information. It carries out research into accidents, safety procedures and other health, safety and welfare topics, and enforces regulations.

Throughout this chapter, many references are made to the HSE website, and it is recommended that you become familiar with this resource – the information provided by the HSE is designed to be easy to follow. Many of the activities in this chapter will require you to access and download information from the HSE website (www.hse.gov.uk/construction/).

The accident statistics recorded by the HSE state that in 2018 and 2019, falls from height were the most common cause of fatalities in the construction industry, accounting for just under half (49 per cent) of fatal injuries to operatives. Many of these incidents could have been avoided by taking precautions.

Effective training in health and safety at the start of your career will embed good practice and awareness of risks. This will go a long way towards improving safety across the industry.

Too often people think that accidents will not happen to them. These may be the very people who do have accidents, though, because they do not give enough thought to how to ensure safe working practice.

All work at height must be properly planned and organised to ensure safe working practice. To help you achieve this, the HSE provides a wealth of knowledge and guidance to make sure that the work complies with the Work at Height Regulations 2005 (as amended).

INDUSTRY TIP

Over half of all deaths during work at height involve falls from ladders, scaffolds, working platforms or roof edges, or through fragile roofs or roof lights. Don't become another statistic – ensure that you properly assess risks and take a safe approach.

Work at Height Regulations (2005) (as amended)

The Work at Height Regulations 2005 (as amended) aim to prevent deaths and injuries caused each year by falls at work. They apply to all work at height where it is likely that someone will be injured if they fall.

The Regulations apply to all work at height where there is risk of a fall that is likely to cause personal injury. They place duties on employers, the self-employed and any person who controls the work of others (such as facilities managers or building owners who may contract others to work at height).

Those with duties under the Regulations must ensure that:
- all work at height is properly planned and organised
- those involved in work at height are competent
- the risks from work at height are assessed, and appropriate work equipment is selected and used
- the risks of working on or near fragile surfaces are properly managed
- the equipment used for work at height is properly inspected and maintained.

For managing work at height and selecting the most appropriate equipment, dutyholders must:

- avoid work at height where possible, for example doing the work from ground level using extending equipment
- where work at height cannot be avoided, use work equipment or other measures to prevent falls, for example cherry pickers or scaffolding
- use work equipment or other measures to minimise the distance and consequences of potential falls, where the risk cannot be eliminated, for example nets or air bags.

For more information, go to the HSE website (www. hse.gov.uk/work-at-height/the-law.htm) where you can access full details of the Work at Height Regulations 2005 (as amended) and download 'Working at height – A brief guide').

Health and safety in construction

Chapter 1 explained legislation relating to the wider aspects of the construction industry and covered health and safety law as well as legislation relating to planning and building control. Legislation that particularly applies when carrying out work at height is briefly described below.

Reporting of Injuries, Diseases and Dangerous Occurrences Regulations (RIDDOR)

Under RIDDOR, fatal and certain non-fatal injuries to workers and members of the public must be reported by employers to the Health and Safety Executive.

Health and Safety at Work Act 1974

All workers have a right to work in places where risks to their health and safety are properly controlled. Health and safety is about stopping you getting hurt at work or becoming ill through work. Your employer is responsible for health and safety, but you have a duty to work safely to protect yourself and those around you.

Provision and Use of Work Equipment Regulations 1998

These Regulations (often abbreviated to PUWER) place duties on people and companies who own, operate or have control over work equipment. PUWER also covers businesses and organisations whose employees use work equipment, whether owned by them or not.

Construction (Design and Management) Regulations 2015

Virtually everyone involved in a construction project has legal duties under CDM 2015 (see Chapter 1 for a full list of dutyholders). The examples shown below relate to two dutyholders – a contractor and a worker.

- **Contractor:** An individual or business in charge of carrying out construction work (e.g. building, altering, maintaining or demolishing). Anyone who manages this work or directly employs or engages construction workers is a contractor. Their main duty is to plan, manage and monitor the work under their control in a way that ensures the health and safety of anyone it might affect (including members of the public). On projects with more than one contractor, contractors work under the control of the principal contractor.
- **Worker:** An individual who actually carries out the work involved in building, altering, maintaining or demolishing buildings or structures. Workers include plumbers, electricians, scaffolders, painters, decorators, steel erectors and labourers, as well as supervisors like forepersons and chargehands. Their duties include co-operating with their employer and other dutyholders and reporting anything they see that might endanger the health and safety of themselves or others. Workers must be consulted on matters affecting their health, safety and welfare.

Personal Protective Equipment at Work Regulations 1992 (as amended)

Employers have duties concerning the provision and use of personal protective equipment (PPE) at work. Information on PPE can be downloaded from the HSE website, which also explains what you need to do to meet the requirements of the Personal Protective Equipment at Work Regulations 1992 (as amended). Typical PPE when working with access equipment includes safety helmet, gloves, overalls and safety footwear. Additional equipment may be required depending on the tasks.

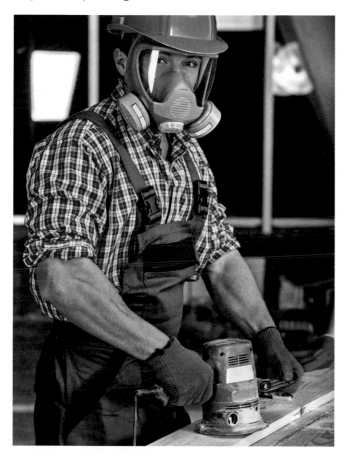

▲ Figure 2.2 Construction worker wearing PPE

Manual Handling Operations Regulations 1992

Employers are required to ensure their employees avoid manual handling where possible, if there is a possibility of injury. If manual handling cannot be avoided, the risk of injury must be reduced by undertaking a risk assessment and ensuring employees follow its recommendations. See the HSE website for more details of risk assessments.

Safe lifting technique

If you cannot use a machine, it is important that you keep the correct posture when lifting any load. The correct technique to do this is known as kinetic lifting. Always lift with your back straight, elbows in, knees bent and your feet slightly apart (see Chapter 1, page 29 for more details on kinetic lifting).

INDUSTRY TIP

Heavy objects that cannot easily be lifted by mechanical methods can be lifted by several people. It is important that one person in the team is in charge and that lifting is done in a co-operative way. It has been known for one person to fall down and the others to then drop the item, particularly when walking backwards or up and down steps.

ACTIVITY

Download 'Manual handling at work – A brief guide' from the HSE website. How does the guide suggest that the risk of injury when manual handling can be controlled?

Risk assessments and method statements

The workforce should be properly trained in the health and safety precautions needed when working at height. Within the construction industry, risk assessments and method statements are used to help manage the work and communicate what is required to all those involved.

Key considerations for all work at height are:
- risk assessments
- precautions required
- method statements.

Risk assessments

Risk assessments are described in detail in Chapter 1 as well as other chapters where risks need to be managed.

Undertaking a risk assessment is a simple but necessary process that must be done in order for you to carry out work and use equipment and materials required for a job safely. It involves looking at the planned work and identifying whether it could result in harm to people. You can then decide whether you have taken enough precautions, or whether more should be done to prevent harm. The purpose of a risk assessment is to reduce both minor and major injuries and fatalities.

Method statement

A method statement is a useful way of recording the hazards involved in specific work at height tasks and communicating the risk and precautions required to all those involved in the work. The statement does not need to be a long document, but it must be clear – it is important to avoid ambiguities or generalisations as these could lead to confusion. Method statements are for the benefit of those carrying out the work and their immediate supervisors and should not be overcomplicated. Sketches to illustrate points can be used where appropriate.

All equipment needed for safe working should be clearly identified and available before work starts. Workers should know what to do if the work method needs to be changed.

Method statements should include the following information:

- **Personal protective equipment**: Describes in detail any PPE required for the task (see page 29).
- **Planning**: Describes potential hazards and information that can be used to make safe decisions before beginning the task. This section includes topics such as site assessments, correct equipment choices, time and resource planning, obtaining

information from qualified persons, obtaining permits, notifying authorities, etc.

- **Preparation**: Provides more site-specific information. While the planning section looks at the bigger picture, the preparation section focuses on what is needed locally at a specific time.
- **Pre-operational inspection**: Includes checks that all equipment to be used is in a safe condition, for example machinery, tools, lifting equipment and associated slings and other such necessary items.
- **Operation**: Outlines the task in sequence. All risks must be identified.
- **Maintenance**: Highlights maintenance regimes or inspection requirements where they are legislated.
- **Emergency procedures**: Highlights essential emergency information. This can include specific first aid procedures if relevant.

Planning that is linked to risk assessments and method statements will go a long way to ensure that a safe preventative approach is taken to all work that you are doing, and this is of paramount importance when using access equipment.

Prepare to use access equipment and working platforms

The most important safety precaution when using access equipment and working platforms is to prepare correctly. Plan to ensure that the correct equipment is selected, the correct personal protective equipment (PPE) is being worn (see page 29) and that adequate consideration has been given to factors such as ground conditions, weather conditions and the height, type and duration of work.

A range of factors need to be taken into account before the final selection of access equipment. Think about the need to prevent falls by selecting the right equipment, and the need to minimise the impact of a fall. These aspects will need to be included in the risk assessment.

▲ Figure 2.3 Working on a ceiling with guard rail protection

INDUSTRY TIP

If you are still in training, you should be supervised by a fully trained and competent person. You should not attempt to carry out any work at height in unsuitable weather conditions (e.g. wind, rain or ice).

Ground conditions and internal/external locations

Whether the work is to be carried out inside or outside, ground conditions need to be level, firm and preferably clean and dry. Adequate steps and additional accessories may need to be used to achieve some of these key points, particularly when working externally. Whatever steps are taken to achieve the required conditions, the method must be safe.

Internal locations

Make sure areas are clear from any potential trip hazards or obstructions. If working on sloping areas or stairs, make sure that the equipment chosen is fit for purpose.

External locations

Soft, rough or uneven ground, or unstable ground conditions will need to be considered. Make sure the site area is clear and tidy and that any likely obstructions or restrictions on full use of equipment are noted.

Make sure that any form of ground support used under any of the selected equipment will not sink into soft ground. For example, when levelling a ladder on a slope, ensure that the wedges or blocks used cannot slip when in use. Of course, the ladder must always be secured, preferably at the top, should anything occur to further affect its stability.

INDUSTRY TIP

Always check that the surface conditions under your access equipment are level, firm, stable and, as far as possible, clean and dry. Make sure stability is not compromised when working on rough ground conditions. This will be particularly important if using tower scaffolds; you will need to check that the structure is level and re-levelled after every move.

Height, type and duration of work

An assessment relative to the height of the project needs to be carried out, irrespective of the actual height. Remember, any height from which a person may fall is deemed to be 'working at height'. You also need to consider the type and duration of work.

Weight

An assessment must be made in relation to the weight of the loads that may be loaded onto the access equipment selected. Most decorating activities do not involve great weight but this should nevertheless be included in the assessment.

Number of operatives

It is important to consider the number of operatives that may be required to carry out the activity. For example, stepladders should be used by only one person and most working platforms built from trestles and lightweight stagings are designed to carry only two people. On large contracts it may be advisable to consider either tower scaffolds or fixed independent scaffolds to accommodate a greater number of personnel.

Weather conditions

All types of equipment can be affected by weather conditions, particularly when used externally. Major problems can arise from using access equipment in very windy conditions, especially if items are not securely tied or materials that might be dislodged are stored on the platform.

Safe working practice will be described by the product manufacturer and is stipulated by regulations such as the Work at Height Regulations 2005 (as amended) and the Provision and Use of Work Equipment Regulations 1998 (PUWER).

▲ Figure 2.4 Take care when using access equipment in adverse weather conditions

Access and egress

Means by which to **access** and **egress** the equipment should be planned before erection. Trying to work around obstructions can lead to hazardous working conditions, so it is best to try and clear working space for the equipment. Make sure that there are clear indicators at the base of access equipment to stop vehicles striking it – traffic cones or barriers work best.

Consideration should also be given to the general public, as it is extremely important to ensure they are protected. Scaffolds and access equipment should have barriers erected around them to prevent people colliding with the equipment or having objects fall on them. If this cannot be fully achieved then, in the case of scaffold, you will need to provide adequate walkways through the scaffold with highly visible standards and debris netting. Figure 2.4 shows a scaffold erected in a busy street. This type of structure will require a temporary structure licence from the local authority. The licence will set out how long the scaffold will be in place and what arrangements must be made for any pedestrian diversions.

> ### KEY TERMS
>
> **Access:** the means to enter or gain entry to a place.
> **Egress:** the means to leave or exit a place.

Types of access equipment and working platforms

This section will look at the various types of access equipment available for carrying out work at height.

Ladders

There are four types of ladders in common use on construction sites, usually for external activities. Occasionally an extension ladder may be used for interior work such as staircases. Remember that leaning ladders are intended only for short-term work.

Pole ladders

A pole ladder is a single-section ladder. Traditionally the **stiles** (the vertical parts) are made from a single tree cut vertically down the middle. Timber ladders have now often been replaced by aluminium ones. Pole ladders are typically used as ladders to access fixed scaffolding erected to the building by qualified scaffolders.

▲ Figure 2.5 A pole ladder

▲ Figure 2.6 A standing ladder

Standing ladders

A standing ladder is a fixed single-section ladder manufactured to length as required, up to a maximum of 4 m, and is commonly used to access scaffold platforms. It has limited flexibility of use due to its being a fixed length, as opposed to a double or triple extension ladder.

Double and treble extending ladders

These consist of two or three sections of ladder connected together by brackets and guides. Many longer extending ladders, in particular triple extending ladders, will be rope operated to enable them to be more easily raised to the operating height required.

INDUSTRY TIP

Timber ladders must not be painted, as this may hide defects. You can use a clear varnish to provide protection if required. Aluminium ladders must not be used near overhead electrical power lines due to the potential for electrical shock.

▲ Figure 2.7 Extending ladder

Roof ladders

Working on roofs of any type is an additional challenge to the decorator, as you may have to carry out maintenance on items such as wooden skylights or decorate whole sections of corrugated iron roof cladding.

Roof ladders are used to gain access over sloping and fragile roof surfaces to reach other items that require decoration. The hook should be used to provide a secure attachment to the roof ridge. Roof ladders should be long enough to span the roof supports (at least three rafters) and be securely placed. However, for increased safety, roof ladders should be used with the guard rail edge protection features shown in Figure 2.8 and only used for short-duration work where possible.

The roof ladder anchorage or hook should rest on the opposite roof and not rely on the ridge tiles for support, as these can easily break away (see Figure 2.9). It is advisable to attach a safety harness and lanyard (Figure 2.10) to the ladder when working from a roof. You should also protect the edges of the roof area where you are working. For more information about the correct PPE see page 84.

▲ Figure 2.10 Man with hi vis and harness

The anchorage or hook goes over the ridge of the roof on to the other side

▲ Figure 2.8 A typical roof ladder

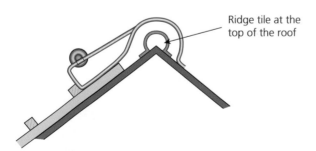

Ridge tile at the top of the roof

▲ Figure 2.9 Roof ladder anchorage or hook resting on the opposite roof

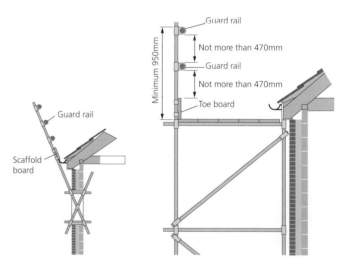

Guard rail

Not more than 470mm

Guard rail

Not more than 470mm

Minimum 950mm

Toe board

Guard rail

Scaffold board

▲ Figure 2.11 Types of edge protection for work on roofs

Ladder classification

Ladders can be made from wood, aluminium, steel or fibreglass.

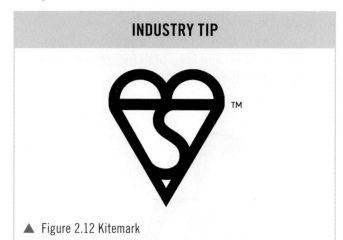
Many accidents happen as a result of the wrong class of ladder being used in working situations. Always aim to use Professional equipment where possible to ensure you are adequately covered for insurance purposes.

Check that the ladder you are using conforms to relevant standards. All ladders should conform to EN131.

The standards that have been revised are designed to ensure ladders are wider, stronger and sturdier. All ladders are designed to carry 150 kg in weight.

There are now two classes:
- Professional
- Non-Professional

▲ Figure 2.13 Ladder classification

The ladder should be marked with the class number, maximum weight and instructions on how to use the ladder safely – make sure you follow these instructions at all times. Using ladders without the BSI Kitemark could mean using equipment that has not been properly assessed against the correct standards. Equipment without the Kitemark might have insufficient strength or be of an incorrect design.

Leaning ladders should be used for:
- short-duration work (maximum 30 minutes)
- light work (up to 10 kg).

Safety precautions for using a leaning ladder:
- Maintain a ladder angle of 75° – remember the '1 in 4 rule' (1 unit out for every 4 units up).
- Do not work from the top three rungs – these provide a handhold.
- Always grip the ladder when climbing.
- Do not over-reach – make sure your body stays within the stiles and keep both feet on the same rung or step throughout the task.

▲ Figure 2.14 Ladder with Kitemark labels

Stepladders

Stepladders are typically used for interior work, or external work where there is a level and solid base. The most common use is for domestic decoration where the work carried out is not excessively high. There are two main types of stepladder:

- with platform steps
- with swingback steps.

Stepladders can be made from timber, aluminium or fibreglass. Fibreglass steps may be chosen for work near electrical installations because they contain no metal parts so there is no risk of electric shock.

Platform steps are particularly useful because they provide a platform to keep materials and tools in easy reach while working at a safe height from the steps.

Swingback steps (see Figure 2.16) are rated Class 1 and are Kitemarked to show that they meet the appropriate standard. Timber steps are made from straight-grained softwood for greater strength and usually have grooved non-slip **treads**. Metal tie rods are fitted under some of the treads to add stability and to stop the stiles springing apart.

> **KEY TERM**
>
> **Tread:** section of steps that the feet stand on.

Stepladders should be used for:

- short-duration work (maximum 30 minutes)
- light work (up to 10 kg).

Safety precautions for using a stepladder:

- Do not work from the top two steps (or the top three steps for swingback/double-sided stepladders) unless you have a safe handhold on the steps.
- Avoid side-on working.
- Do not over-reach – make sure your body stays within the stiles and keep both feet on the same rung or step throughout the task.

▲ Figure 2.16 Swingback steps

▲ Figure 2.15 Fibreglass platform stepladder

Trestles

Trestles are made from timber, aluminium and fibreglass and create an A-frame shape when in use. Trestles are usually tapered towards the top and should be wide enough to take two 230 mm scaffold boards or one lightweight 450 mm-wide staging.

Each side of the trestle should have at least two tie bars. The trestle is designed so that it does not collapse when it is opened, thanks to a special hinge. When the hinge is fully opened, the stiles lock against one another.

When using a trestle with a working platform, your risk assessment will determine whether you need to use guard rails (see page 82) and a **toe board** (see Figure 2.19). You should access the platform by an additional ladder. Never use trestles as steps – the space between supports is designed to allow variation in platform height and is too wide for stepping.

▲ Figure 2.17 Timber trestle

Adjustable-height steel trestles

Adjustable-height steel trestles are commonly used by bricklayers and plasterers together with scaffold boards to provide low-level platforms. They can also be useful for low-level ceiling work by painters and are best used with a handrail attachment to reduce the risk of falling. They are sometimes referred to as **bandstands**.

Adjustable-height steel trestles are designed to be used with four standard 225 mm-wide scaffold boards or two 450 mm-wide lightweight stagings. These are specially constructed timber and aluminium platforms designed to span greater widths than scaffold boards. The trestles feature an adjustment pin that is permanently secured, therefore reducing the cost of replacing lost pins and wires.

It is best to use a purpose-made handrail system in order to fully comply with the Work at Height Regulations 2005 (as amended), but this depends on your risk assessment for the particular task at hand.

Some manufacturers offer purpose-made handrail and toe-board systems that can be attached either to the steel trestles themselves or to the working platform.

▲ Figure 2.18 Adjustable-height steel trestles

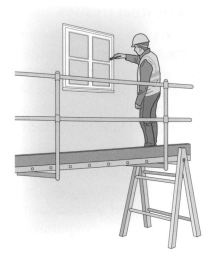

▲ Figure 2.19 Handrails and toe boards are used to comply with the Work at Height Regulations 2005 (as amended)

Podium steps

Podium steps, or podiums, provide low-level height access and offer a firm platform with adjustable height, as well as a guard rail. The steps may be lightweight, self-erecting or folded prior to erection, to enable them to pass through standard doors and corridors. Figure 2.20 shows a fully erected podium.

▲ Figure 2.20 Podium steps

A key message from the HSE is to put tools and materials onto the podium at ground level where possible and always to close and lock the gate before starting work.

Scaffold boards

Scaffold boards (planks) are used to provide a working platform on trestles and tubular scaffolds. If you are using scaffold boards as a working platform you should always consider whether a safer mode of scaffolding could be used instead. Painters in particular use two boards, and this is suitable only in certain situations such as working from low-level platforms using hop-ups (see page 73). For some short-term jobs such as papering ceilings it may be better to use this type of working platform, provided there is sufficient support over the length. This will mean supports every 1.2 m. Where possible, a lightweight staging should be used instead.

There are four safety points you should look for when inspecting scaffold boards before use. The boards must:

- be straight grained
- be free from knots and splits in the timber
- be free from any decay (usually at the ends)
- have end bands in place to protect the end grain.

▲ Figure 2.21 Scaffold boards in use

Lightweight staging

Lightweight staging is a specially constructed timber and aluminium platform designed to span greater widths than scaffold boards. Lightweight stagings can be used without intermediate supports when used on trestles.

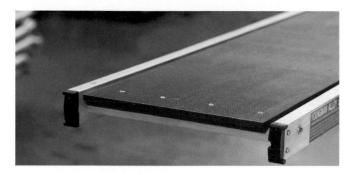

▲ Figure 2.22 Lightweight staging

Sizes of stagings vary, starting at 450 mm wide, with lengths of 1.8 m to 7.3 m. Stiles are reinforced with high tensile steel wire. Cross supports are every 380 mm or 450 mm along the length of the staging and can be reinforced with steel ties.

> **INDUSTRY TIP**
>
> Stagings are designed to take a maximum load of three people, together with lightweight equipment.

Folding work platform or hop-up

Folding work platforms, also known as hop-ups, generally have a 600 × 600 mm square platform – this is now the preferred choice of health and safety officials and is often the only size allowed on sites. Typically, they are built of aluminium alloy and will withstand loads of 175 kg. The fold-flat models are easily folded for moving and transporting.

▲ Figure 2.23 Hop-up or folding work platform

This type of equipment can be used safely with lightweight platforms or scaffold boards to form a low-level working platform ideal for decorating or papering ceilings. It does not take up too much space and can be easily transported when moving to new contracts.

Tower scaffolds

An aluminium alloy tower scaffold is often chosen by decorators for many interior and exterior work situations. It may be static or mobile, and this will be determined by the work height, type of activity and the duration of the work.

▲ Figure 2.24 A mobile tower scaffold

Other accessories for use with access equipment

Sometimes it may be necessary to use another accessory or device with the piece of access equipment.

A sturdy aluminium stand-off (see Figure 2.25) fits directly onto your ladder and creates a 300 mm gap between the ladder and wall at the top. It reduces the risk of the ladder twisting or slipping and prevents damage to guttering. It is essential that ladders are not rested on guttering – apart from causing damage to the guttering, they are more likely to slip.

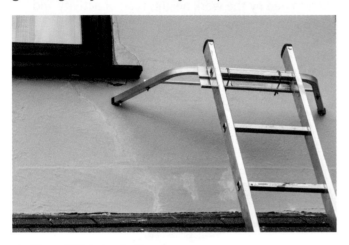

▲ Figure 2.25 A stand-off accessory

▲ Figure 2.26 Proprietary ladder stabiliser

A proprietary ladder stabiliser (Figure 2.26) is used at the base of a ladder to effectively widen the base, making it less likely to slip.

<table>
<tr><td>

INDUSTRY TIPS

Do not use a gutter to support any ladder. It is better to use a ladder stand-off accessory (see Figure 2.26) to keep the ladder off the gutter when working.

It is always best to combine accessory devices with securing the ladder at the top for added safety.

</td></tr>
</table>

Keeping equipment on site

Whether equipment is hired or owned, there are a number of key points to incorporate into your planning. Make sure only the required amount of equipment is delivered and kept on site during operations, as this can affect costs and impact greatly on the original estimate. Unused equipment may be stolen or damaged and must also be stored. Ladders present particular additional risks as they can be taken and used for burglary, so whenever possible equipment should either be stored in locked compounds or chained together with padlocks. The cost of insuring equipment will increase with the amount of time it is left on site; make sure there is adequate insurance cover for loss or damage. Finally, equipment can add greatly to the overall cost of a job if it is no longer required but still on hire.

▲ Figure 2.27 Make sure equipment kept on site is safe and secure

Guidance information

When preparing access equipment and working platforms for use it is essential to follow all the guidance information available, and this includes manufacturers' instructions, the Work at Height Regulations 2005 (as amended) and the Provision and Use of Work Equipment Regulations 1998 (PUWER).

If you are going to work at height you should be trained and competent and should be able to complete the task safely and operate the selected access equipment.

The law requires employers and self-employed contractors to assess the risk of working at height and to organise and plan the work so it is carried out safely.

Manufacturers' instructions

It is extremely important to obtain guidance from either the manufacturer or the hire company to enable you to comply with their recommendations. It should be possible to obtain detailed information for any type of equipment you use. Figure 2.28 shows part of an instruction manual for a mobile access tower.

▲ Figure 2.28 An example of an instruction manual for a mobile access tower

ACTIVITY

Look online for an example of a manufacturer's instructions for equipment for working at height. What items help to prevent falls from the items of equipment shown in the manufacturer's leaflet?

You may need to study the Work at Height Regulations 2005 (as amended) for clear guidance on the requirements for guard rails, toe boards, barriers and similar collective means of protection as defined in Schedule 2. Go to: www.legislation.gov.uk/uksi/2005/735/schedule/2/made

Inspection of access equipment

Your employer is ultimately responsible for ensuring that the equipment supplied to you is safe and fit for purpose, but remember that as the user you are responsible for your own safety. It is important that visual checks are carried out on a regular basis to ensure that the equipment you are using is still fit for use.

Some of the information given in this section will help you carry out systemised checks and inspections in line with the regulations. Always report any defects found during your daily checks so that action may be taken to maintain everyone's safety.

▲ Figure 2.29 Ladder tag inspection record

Inspection time periods

Access equipment should be checked:

- pre-erection
- post-erection
- before handing over
- if there are poor weather conditions
- if major alterations have been made
- every seven days
- post-accident/post-incident, as the condition of the equipment may have altered.

Procedure for carrying out visual checks on access equipment prior to use

Before using access equipment, a visual safety check of the equipment should be made and the correct paperwork completed. While you are training, you should be overseen by a competent person who can provide you with the correct guidance to complete these simple documents, while identifying any potential hazards.

Schedule 7 of the Work at Height Regulations 2005 (as amended) requires the following particulars to be recorded:

1 the name and address of the person for whom the inspection was carried out
2 the location of the work equipment inspected
3 a description of the work equipment inspected
4 the date and time of the inspection
5 details of any matter identified that could lead to a risk to the health or safety of any person
6 details of any action taken as a result of any matter identified in point 5
7 details of any further action considered necessary
8 the name and position of the person making the report.

Checks and inspections of ladders, steps and trestles

Ladders

The areas to be checked with regard to ladders are shown in the six-monthly inspection records that are provided for recording this information (see Figure 2.29 on page 75). A basic pre-use checklist (see page 77)

is completed as part of the inspection on a daily basis. This inspection is done to ensure the equipment is being safely used and at the correct angles, on safe ground and the correct PPE is used when required. These steps will help to ensure that there are no areas for concern such as whether the ladder is being used at the correct angle of 75°, or a ratio of one out to four up (see page 79).

Figure 2.30 shows a ladder with labels to indicate the areas that would be checked as part of an inspection record.

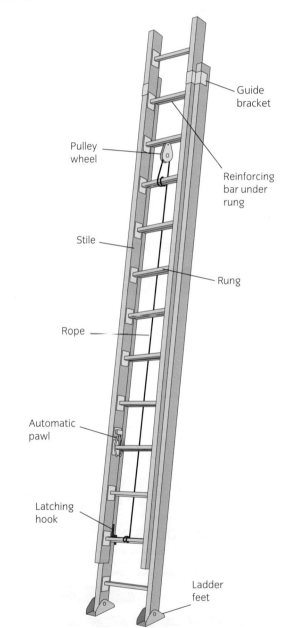

Guide bracket

Pulley wheel

Reinforcing bar under rung

Stile

Rung

Rope

Automatic pawl

Latching hook

Ladder feet

▲ Figure 2.30 Labelled ladder

Below are a number of questions that need to be answered as part of the daily pre-use checklist.

Basic pre-use checklist

Employers and employees and all users of ladders should be able to answer 'Yes' to each of the following questions, or to the alternative given, before a job is started.

	Yes	No
Is a ladder, stepladder, etc., the right equipment for the work?		
If so, is the equipment in good condition and free from slippery substances?		
Can the leaning ladder be secured at the top?		
If not, can it be secured at the bottom?		
If a ladder has to be used and cannot be secured, will a second person stationed at the base provide sufficient safety?		
Does the ladder project above the platform by 1 m?		
Is there an adequate handhold at the place of landing?		
Does the ladder project above the platform by 1 m?		
Is there an adequate handhold at the place of landing?		
Are there platforms at 9 m maximum intervals?		
Is the ladder angle correct?		
Is the support for the ladder adequate at both the upper point of rest and the foot?		
Is the ladder properly positioned?		
If it is necessary to carry tools and equipment, has provision been made for carrying them so that the user can keep their hands free for climbing?		
If an extension ladder is used, is there sufficient overlap between sections?		
On the stepladder, are the locking bars or support ropes in good condition?		
Can the stepladder be placed sufficiently near the work on a firm, level surface?		
Is the ladder clear of overhead electric cables?		

ACTIVITY

Visit the Ladder Association website (https://ladderassociation.org.uk/) and download the ladder safety guide, 'Get a Grip on Ladder Safety'. Using the information provided in the guide, write down the key points regarding stepladder inspection.

Steps and trestles

Figures 2.31 and 2.32 illustrate the parts that will require checking on steps and trestles. The condition can be recorded on the inspection records.

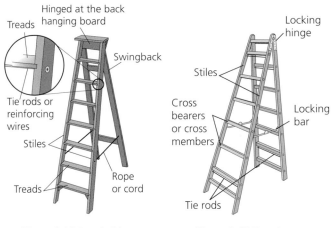

▲ Figure 2.31 Stepladder ▲ Figure 2.32 Trestle

Inspection log

An inspection log is attached to the side of a piece of equipment confirming that it has been inspected, and that it is either safe or unsafe to use. The green log denotes that it has passed the inspection and is safe, while the red log indicates that it has not passed and is not safe.

PASSED ✓
Ladder ID Number:
Allocated To:
Date of Last Inspection:
Date of Next Inspection:
Inspected By:
LADDER SAFETY INSPECTION

FAILED ✗
Ladder ID Number:
Allocated To:
DO NOT USE
Date of Inspection:
Inspected By:
LADDER SAFETY INSPECTION

▲ Figure 2.33 Inspection labels

Below is an example inspection checklist for ladders, steps and trestles.

LADDERS			
(6-MONTHLY INSPECTION)			
Department/Location: Ladder No:			TICK OK
Inspected by:			
NO: ITEM:		Condition	↺
	STRAIGHT LADDER		
	Loose rungs (move by hand)		
	Loose nails, screws, bolts, etc.		
	Loose mounting brackets, etc.		
	Cracked, broken, split stays		
	Splinters on stays or rungs		
	Cracks in metal stays		
	Bent metal stays or rungs		
	Damaged/worn non-slip devices		
	Wobbly		
	STEPLADDER		
	Wobbly		
	Loose/bent hinge spreaders		
	Stop on spreaders broken		
	Loose hinges		
	EXTENSION LADDER		
	Defective extension locks		
	Defective rope pulley		
	Deterioration of rope		
	TRESTLE LADDER		
	Wobbly		
	Defective hinges		
	Defective hinge spreaders		
	Stop on spreads defective		
	Defective centre guide for extension		
	Defective extension locks		
	FIXED LADDER		
	Ladder cage		
	Deterioration in all metal parts		
	GENERAL		
	Painting of wooden ladders		
	Identification		
	Storage		

Checks and inspections on leaning ladders

Set-up

- Do a daily pre-use check (include ladder feet).
- Secure the ladder.
- Ground should be firm and level.
- Maximum safe ground side slope 16° (level the rungs with a suitable device).
- Maximum safe ground back slope 6°.
- Have a strong upper resting point (not plastic guttering).
- Floors should be clean, not slippery.

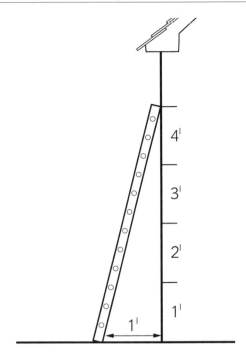

▲ Figure 2.35 The correct angle for a ladder is 1:4

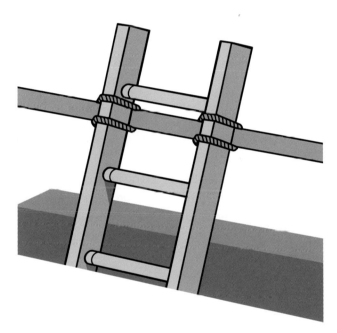

▲ Figure 2.34 The top of a ladder secured by lashing

In use

- Short-duration work (maximum 30 minutes).
- Light work (up to 10 kg).
- Ladder angle 75° – one-in-four rule (one unit out for every four units up).
- Always grip the ladder when climbing.
- Do not over-reach – make sure your belt buckle (navel) stays within the stiles and keep both feet on the same rung or step throughout the task.
- Do not work off the top three rungs – they provide a handhold.
- Always maintain three points of contact with the ladder, preferably both feet and at least one hand.

Checks and inspections on stepladders

Set-up

- Do a daily pre-use check (include stepladder feet).
- Ensure there is space to fully open the stepladder.
- Use any locking devices.
- Ground should be firm and level.
- Floors should be clean, not slippery.

In use

- Short-duration work (maximum 30 minutes).
- Light work (up to 10 kg).
- Do not work off the top two steps (top three steps for swingback/double-sided stepladders) unless you have a safe handhold on the steps.
- Avoid side-on working.
- Do not over-reach – make sure your belt buckle (navel) stays within the stiles and keep both feet on the same rung or step throughout the task.

ACTIVITY

1 When would ladders be suggested as the right equipment for the job?
2 What is the correct angle for the ladder?
3 What are the dangers of using ladders near electric cables?

Checks and inspections on mobile towers

The Work at Height Regulations 2005 (as amended) require that mobile access scaffold towers are inspected by a competent person after assembly and before use. A written report of that inspection must be completed before the competent person goes off duty and a copy of that report must be given to the person for whom the report was completed within 24 hours.

PASMA, in consultation with the HSE, has developed the PASMA Tower Inspection Record for this purpose. As well as providing a visual indicator of the tower's inspection status, it acts as a written report, and by affixing the record to the tower, you satisfy the requirement to give it to the person for whom it was completed within 24 hours.

▲ Figure 2.36 PASMA Tower Inspection Record

> ### KEY TERM
>
> **PASMA:** Prefabricated Access Suppliers' & Manufacturers' Association Ltd – the lead trade association for the mobile access tower industry.

Time periods

All towers must be inspected following assembly and then at suitable regular intervals. In addition, if the tower is used for construction work and a person could fall 2 m or more from the working platform, it must be inspected following assembly and then every seven days. Stop work if the inspection shows that it is not safe to continue and make sure to put right any faults.

HSE guidance states that in addition to the seven-day frequency of inspections, a tower should be inspected after any event likely to have affected its stability or structural integrity, such as adverse weather conditions. You may be able to think of other events that could have such an effect. It is important to note that the regulations do not require a written report each time a tower is moved or relocated to the same site. However, if guard rails or other components have to be removed to enable the tower to be moved past an obstruction, a pre-use check should be undertaken by a trained and competent user to make sure the tower has been reinstated correctly.

Completion of records

When the record is full, it is removed from the tower (if the tower is still being used, a new Tower Inspection Record is commenced) and retained as a record of the inspections until the work is completed, and thereafter at your office for a further three months, as required by the regulations.

Incomplete, damaged and unsafe towers

As well as being suitable for recording inspections, the reverse of the PASMA Tower Inspection Record can also be used as a visual indicator that the tower is not to be used because it is incomplete, damaged or otherwise unsafe. If your tower is incomplete or is in a dangerous condition, you must let other people know by attaching a 'Tower incomplete', 'Tower damaged' or 'Tower unsafe' sign in a prominent position or adjacent to an access point. This will ensure that any potential users are aware of its condition and do not attempt to use it. Always keep up to date with any changes in the regulations.

Inspection

PUWER (Provision and Use of Work Equipment Regulations) 1998 and the Work at Height Regulations 2005 (as amended) make it a legal requirement to ensure that all commercial scaffold towers are safe to use. Figure 2.37 shows an inspection system in use and the **scaff tags** being checked and updated. The scaffold should be inspected on a regular and systematic basis so that it complies with the regulations.

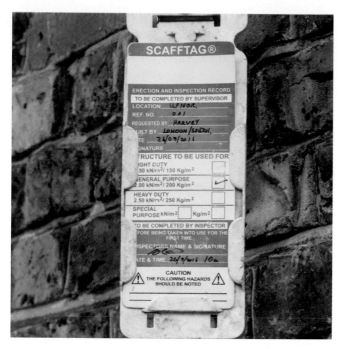

▲ Figure 2.37 Equipment inspection

The requirement for inspection is different for small towers under 2 m, and for towers of 2 m and above.

If the tower is under 2 m in height, the tower must be inspected:

- after assembly in any position
- after any event liable to have affected its stability
- at suitable intervals, depending on frequency and conditions of use.

If the tower is over 2 m in height, the tower must be inspected:

- after assembly in any position
- after any event likely to have affected its stability
- at intervals not exceeding seven days.

Safety checklist

- Ensure all brace claws operate and lock correctly before erecting.
- Inspect components before erecting.
- Inspect tower prior to use.
- Ensure tower is upright and level.
- Ensure that **castors** are locked and legs correctly adjusted.
- Diagonal braces fitted?
- Stabilisers/outriggers fitted as specified?
- Platforms located and windlocks on?
- Toe boards located?
- Check that guard rails are fitted correctly.

Figure 2.38 shows the various parts of a tower scaffold that form part of a visual check during the inspection process. It is very important to ensure that any defective equipment is taken out of use and its condition reported to your line manager or supervisor. Make sure that others who may be likely to use the faulty equipment are informed of its condition.

If items are missing, such as toe boards from tower scaffolds for example, it may be possible to collect these and put them in place. Once again, remember that you need to have been trained to ensure they are correctly fixed.

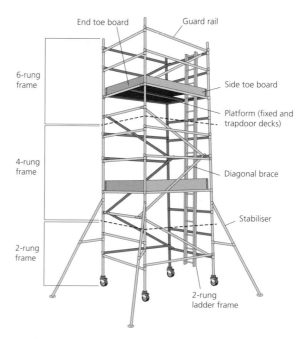

▲ Figure 2.38 Example of a tower scaffold

2 USE ACCESS EQUIPMENT

This section looks at the factors affecting the use of access equipment.

Risk assessments for access equipment and working platforms

Risk assessments were covered earlier in this chapter, but here we will look at how to apply them in the context of working at height.

The Construction, Design and Management (CDM) Regulations 2015 require employers to make a suitable and sufficient assessment of the health and safety risks to employees and non-employees arising from their work activities. Risk assessments and method statements are an essential requirement of any health and safety plan. It is important to focus on the prevention of accidents and ill-health, rather than reacting to incidents after they have happened.

Guard rails

There are specific requirements for protection within Schedule 2 of the Work at Height Regulations 2005 (as amended). The most important of these relate to guard rail locations and dimensions, and Figure 2.39 highlights the key dimensions. The regulations also state that the minimum height of a guard rail should be no less than 950 mm from the working platform.

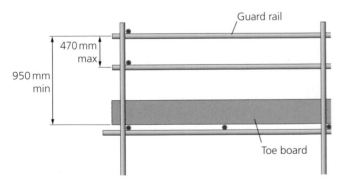

▲ Figure 2.39 Guard rails and toe board

The HSE suggests that the following approaches are taken for work at height:

- Avoid working at height where it is reasonably practicable to do so, and select the most appropriate means to minimise the risk from falls.
- Prevent any person falling a distance liable to cause personal injury, for example by using a scaffold platform with double guard rail and toe boards.
- Arrest a fall with equipment to minimise the distance and consequences of a fall, for example safety nets where work at height cannot be avoided or the fall prevented.

Risk assessments should always address the following:

- task description and location
- expected duration
- hazards identified
- population exposed
- risks arising
- control measures.

All employers are responsible for producing a risk assessment, and if they employ five or more employees, a written form must be provided. Each risk assessment should be written in clear, understandable language and should be explained to workers before the task or process begins. Work at height is a particular area for which risk assessments should be undertaken.

Hazard identification

The following points may be considered when identifying hazards:

- falls from heights (people/materials)
- slips, trips, cuts and abrasions
- faulty equipment
- altered/removed parts
- control measures.

ACTIVITY

Visit the HSE website and download 'Risk assessment: A brief guide to controlling risks in the workplace' (www.hse.gov.uk/pubns/indg163.htm). Use the guide to list the five steps to risk assessment.

The table opposite applies the information shown in the HSE's 'Risk assessment: A brief guide to controlling risks in the workplace' to some specific examples of tasks.

Task	Hazard	Risk	Control
Using ladders as a means of access from one level to another	• Falling from the ladder • Displacement of the ladder • Failure of the ladder	• Major or lost-time injury to head, legs, arms or internal organs	• Inspect the ladder for visible defects before use • Ensure that the ladder is secured to the landing point and, if necessary, at the base • Use both hands when climbing up or coming down the ladder
Working at or adjacent to a leading edge	• A person or persons falling from one level to another • Materials falling from heights	• Fatal, major or lost-time injury • Fatal, major or lost-time injury due to being struck by falling material(s)	• Check the work location prior to commencement • Confirm the positioning, integrity and suitability of the barrier • If considered unsuitable, stop and have the assembly upgraded
Working from bandstand scaffolds or hop-ups	• A person or persons falling from the work platform • Structural failure of the assembly	• Major or lost-time injury	• Authorisation for the use of bandstand scaffolds or hop-ups to be recorded in the Safety Method Statement • The platform height must not exceed 1.2 m • Overloading must be prohibited • Safe ladder access point must be established
Working from steps or ladders	• Failure of the steps/ladder • Falling from the steps/ladder • Over-reaching from the steps/ladder	• Fatal, major or lost-time injury	• Task to be subject to a specific risk assessment • If authorised for short-duration, light-duty tasks only • Ladder to be footed and secured • Lone working prohibited
Working from steps near energised electrical apparatus or apparatus capable of being energised	• Making contact with or causing an arc from the apparatus to the steps	• Electric shock or flashover fatal or major injury • Damage to equipment	• Steps made from non-conductive material only may be used in such locations
Working from or resting ladders/steps against cable trays or in-situ pipework	• Failure of the tray suspension system • Failure of or damage to in-situ pipework	• Fall from height – fatal or major injury • Uncontrolled release of liquid, gas or other substance being piped • Major injury and/or damage	• Working from suspended cable trays prohibited • Resting ladders/steps against in-situ services prohibited unless authorised by the service owner
Issuing or using safety harnesses for the purpose of arresting the fall of a person	• Use of untrained personnel • Failure to inspect the system before issue/use • Attaching the system to an unapproved fixing point	• Fatal, major or lost-time injury in the event of a malfunction of the system or the anchorage point	• Safety harnesses to be issued to and used by trained personnel only • The task and location shall be subject to a specific risk assessment • The system shall be subject to a record-keeping regime as prescribed by the manufacturer

For further information and to view sample risk assessments go to www.hse.gov.uk/risk/casestudies and view the example assessment for a plastering company.

Personal protective equipment (PPE)

Remember that PPE must be worn whenever the work requires it, especially on construction sites. Typically this will be gloves, safety boots, hard hat, overalls and high-visibility jacket. In addition to the basic safety equipment you may also be required to wear goggles and specialist equipment such as safety harnesses with attached lanyards or fall arrest devices. This will have been highlighted within the method statement and will be specific to the task.

Manual handling

When erecting, dismantling or handling access equipment you should ensure that the correct manual handling techniques are employed to minimise injury.

Safe lifting relates to lifting items from the floor, and could apply to lifting ladders, frames or other access equipment from the floor. The correct technique is known as **kinetic lifting**, which means always lifting with your back straight, elbows in, knees bent and your feet slightly apart. The procedure for kinetic lifting is described in Figure 1.43 (page 29).

This lifting technique can also be used when lifting materials onto the work platform, but generally decorators do not need to put down any very heavy materials when working at height. It is important to consider safe working load limits on the scaffold structure or working platform. Do not, for example, overload a tower scaffold with more people than it is designed for. This information will normally be included in the manufacturer's guidance leaflet.

You may need to lift items that are awkwardly shaped, and you will need to ensure that they are correctly balanced. Make an assessment as to whether you are

strong enough to support, carry and handle all the equipment you need to use.

INDUSTRY TIP

Remember that some parts of access equipment are heavy, awkwardly shaped or particularly long, so do call for assistance whenever a task becomes too difficult for one person.

Figures 2.40 to 2.43 demonstrate techniques for handling some items that may not fit the standard categories of manual handling. However, it is also extremely important to be properly trained and to have received a demonstration of the various methods of manual handling before you put them into practice.

▲ Figure 2.40 Carrying a scaffold board

▲ Figure 2.41 Two people carrying a ladder

▲ Figure 2.42 Carrying steps correctly

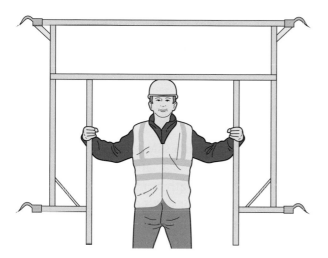

▲ Figure 2.43 Carrying a tower scaffold frame

Storage requirements for access equipment and working platforms

Storage areas for access equipment should ideally be well ventilated, dry and secured or protected from the weather and theft or damage. It is recommended to use properly designed racks to enable ladders, scaffold boards and stagings to be stored flat, horizontally and adequately supported along their length to prevent bowing, warping and twisting. All materials are affected by damp in particular, but heat can also cause problems, especially to timber-based products. Metal items may require hinges, joints and so on to be regularly oiled to keep them in good working order.

▲ Figure 2.44 Store equipment correctly and safely

Repairs

You should not attempt to repair a ladder unless you are qualified to do so. You should seek advice from the manufacturer about repair or replacement.

> **INDUSTRY TIP**
>
> Oiling metal items such hinges or other working parts after use and before storage is a good maintenance tip, as well as checking for damage or decay.

Safety when not in use

When ladders are not in use, such as at night or weekends, it is necessary to prevent anyone climbing the scaffold. This is not just for security purposes (i.e. to prevent break-ins), but also to ensure safety. If ladders are left 'open' during non-working periods and someone does have an accident on or around the scaffold, the company would be liable (i.e. responsible by law) for not making the site safe.

▲ Figure 2.45 Horizontal storage rack

IMPROVE YOUR ENGLISH

Visit the following ladder safety website: www.ladders-999.co.uk/ladder-safety#Storing_Ladders and list the correct storage methods shown under 'Step 6'.

Use access equipment

The erection of ladders and scaffold towers is shown in the step-by-step instructions on page 87 and pages 88–89, but the principles can be applied to all types of access equipment. Ensure that there is a fall protection device (such as a safety harness and lanyard) or procedure in place at all times when dismantling such equipment.

▲ Figure 2.46 Safety harness

This is a full harness made from nylon that is provided to enable a safety lanyard to be attached. It provides a means of fall protection when working on projects such as bridges or high roof structures.

▲ Figure 2.47 Lanyard or safety line and clip

A lanyard is attached to a safety harness as well as to a secure anchor point and a safety rope. This safety rope can be of fixed length or may incorporate an **inertia-operated anchor device** that locks to prevent a fall when weight is applied to it. This equipment will typically be worn as part of a fall protection system.

KEY TERM

Inertia-operated anchor device: a safety device attached to a safety line that works in the same way as a car seat belt to lock when weight falls on it.

Erecting and lowering ladders

The sequence of erecting and moving ladders is shown below. If possible, you should take the ladder down and carry it horizontally, at your side. This will help you to keep your balance and make the ladder easier to handle. Lowering the ladder will be a reverse procedure of the method for putting it up.

Step-by-step process for erecting and moving a ladder

STEP 1 When setting up the ladder for use, push the base of the ladder into the bottom of the wall and start lifting as shown. Ensure that your back is kept straight, and bend from the knees.

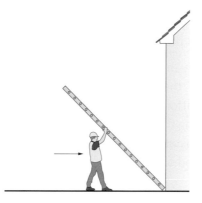

STEP 2 Continue to push the ladder into an upright position. Ensure that you only lift a weight that you are able to hold.

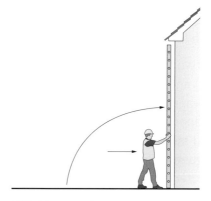

STEP 3 Continue lifting until the ladder is in a near-vertical position, and start to pull the bottom of the ladder out while the ladder is resting on the wall.

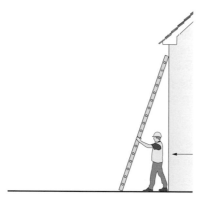

STEP 4 Once the ladder is in an inclined position it may be easier to move to the back of the ladder to set the correct angle.

STEP 5 Once you have finished using the ladder in one position, you may wish to move it a short distance to another position. This can be achieved by holding the ladder against your shoulder to achieve balance. Your hands will be placed one higher and one lower to enable lifting and to maintain balance. This method should only be used when moving short distances.

INDUSTRY TIP

If you cannot comfortably carry the ladder by yourself, do not risk an accident. Ask for help instead.

Erecting and dismantling tower scaffolds

BS EN1004-designed tower systems

BS EN1004 is the European product standard for mobile access (scaffold) towers. Towers that conform to this standard can be built using one of two recognised safe systems of building and dismantling scaffold towers. The key rule in both systems is that you must never stand on an unprotected platform – that is a platform which is not fitted with guard rails around all the edges.

The two safe systems are called 3T ('through-the-trap') and AGR (advance guard rail). With 3T, the user is always in a protected position within a platform trapdoor opening when installing or removing the guard rails. AGR systems allow the user to install the guard rails in advance so they never stand on an unprotected platform.

Always assemble and dismantle the tower from a safe and secure position on a platform with guard rails. You should never stand on the frames of a tower when assembling and dismantling.

Stabilisers

BS EN1004 towers have stabilisers supplied as part of the tower, which are not optional. Mobile towers must be fitted with stabilisers when used freestanding.

Fitting stabilisers on mobile towers at 2.5 m height or less

It depends on the manufacturer's instructions for the tower height you are building as to whether shorter towers will require stabilisers. The old three-to-one rule for the ratio of tower height to base dimension no longer works for determining the stability of a mobile tower: the base dimensions are now determined by a complex calculation in EN1004 which considers many factors. The only way you can determine if stabilisers are required and which size of stabilisers to use is to look at the manufacturer's instructions and the schedule of components. Remember that you must fit stabilisers at the first opportunity in the build sequence and remove them at the end when dismantling.

Using a tower

Never use a tower:

- in strong winds
- as a support for ladders, trestles or other access equipment
- with broken or missing parts
- with incompatible components.

Moving a tower

When moving a tower always:

- reduce the height to a maximum of 4 m
- check that there are no power lines or other obstructions overhead
- check that the ground is firm, level and free from potholes
- push or pull using manual effort from the base only.

Never move a tower while people or materials are on the tower, or in windy conditions.

IMPROVE YOUR MATHS

Using a reliable internet source such as www.hss.com/hire, provide a cost for hiring a tower scaffold for a week and compare with the cost of buying a similar item using BOSS access equipment. BOSS does not sell direct to the public so use a local stockist to source the typical cost of buying.

The step-by-step process opposite shows a shortened sequence for the erection of a tower scaffold from a manufacturer's guidance leaflet. Before erecting a tower scaffold, it is advisable that you attend a training course run on behalf of PASMA, at the end of which a certificate of competence will be issued. Whether equipment is hired or owned, it is important to follow the manufacturer's instructions.

In each illustration, the parts coloured red are the parts discussed in the instructions. You may require an assistant to hand you the various parts of the tower as it gets taller.

Step-by-step process for erecting a tower system

STEP 1 Push the wheels into position and lock.

STEP 2 Holding the frame upright, attach the horizontal cross brace. Ensure that this is locked in position. Note how the brace locks onto the inside of the frame. Repeat this activity by locking the brace onto the other side of the frame. Note that this is locked onto the rung of the frame. Attach the brace from one frame to the other to form a solid base for the tower.

STEP 3 Attach the first set of frames to each side. Attach diagonal braces as shown in the diagram. Ensure that the interlock clips are locked in place.

STEP 4 Level the tower using a spirit level. You may need to adjust the legs to obtain the correct level. Attach outriggers/stabilisers to each side of the tower.

STEP 5 Attach the bottom platform first, then attach the platform with the trapdoor. Climb up the internal ladder, push open the trapdoor and attach guard rails to both sides before stepping through the trapdoor and standing on the platform.

STEP 6 Attach two more diagonal braces as shown, and insert the next two tower frames while standing on the platform.

STEP 7 Attach two more diagonal braces as shown, to stabilise the upper frames. Attach the next trapdoor platform above your head and add in a further diagonal brace.

STEP 8 As before, climb up the internal ladder, push open the trapdoor and attach guard rails on both sides before stepping through the trapdoor and standing on the platform.

STEP 9 Attach toe boards around the top platform. Position the tower in its working position and lock the brakes before re-checking the tower is level.

IMPROVE YOUR ENGLISH

It is possible to hire or purchase outright a tower scaffold. Explain the advantages and disadvantages of each of these two options.

ACTIVITY

Go to Schedule 2 of the Work at Height Regulations 2005 (as amended) (www.legislation.gov.uk/uksi/2005/735/schedule/2/made). This covers guard rails, toe boards and other types of barriers required when working at height.

Look at the requirements shown under item 3 of Schedule 2. What are the requirements in relation to the height of handrails?

Test your knowledge

1 Which item would provide the most protection against cuts and abrasions?

   a Safety gloves

   b Hi vis jacket

   c Safety goggles

   d Dust mask

2 Working at height is considered to be:

   a 1 m above ground

   b 2 m above ground

   c Any height above ground

   d 3 m above ground.

3 What is the **most** important reason why a ladder should not be painted?

   a It will make it slippery

   b The paint may hide defects

   c The paint will come off on you

   d It will need repainting

4 Tower scaffolds are suitable for:

   a interior use only, with stabilisers

   b exterior use only, with stabilisers

   c working at low level heights

   d both interior and exterior use.

5 When using a tower scaffold, it is found that the wheel brakes do not work. What should be done?

   a Only use the tower if the floor is level.

   b Get someone to hold the tower while it is in use.

   c Do not use the tower until the brakes are fixed.

   d Wedge the wheels so they do not move.

6 To reach the working platform of a tower scaffold it is necessary to climb up:

   a a ladder on the outside of the tower

   b the ladder built inside the tower

   c a ladder leant against the side

   d the outside of the bracing.

7 The gap between the intermediate and top guard rail should **not** exceed:

   a 950 mm

   b 400 mm

   c 470 mm

   d 520 mm.

8 The **minimum** height of the top guard rail on access equipment is:

   a 950 mm

   b 900 mm

   c 850 mm

   d 800 mm.

9 Which of the following describes the **correct** manual handling procedure when lifting materials or equipment?

   a Back bent, knees bent, feet together

   b Back straight, knees bent, feet slightly apart

   c Back straight, knees straight, feet together

   d Back bent, knees bent, feet slightly apart

10 Which organisation provides good practice guidance and certification for training in the use of mobile scaffolds?

   a Ladder Association

   b HSE

   c NASC

   d PASMA

11 The aim of the Work at Height Regulations 2005 (as amended) is to:

   a prohibit the use of long ladders

   b stop people working at height

   c prevent deaths and injuries

   d make people wear PPE.

12 Which set of conditions is **best** for storing scaffold components?

   a Cool and dry

   b Warm and moist

   c Warm and dry

   d Cool and damp

Practical task

Prepare, erect and dismantle a mobile tower scaffold

This activity should be carried out in pairs. It is to be carried out on two separate occasions with a different person leading each time, in order to assess individual performance. Ensure that the manufacturer's guidance for erecting and dismantling are made available before the start of this activity.

The tasks to be completed are as follows:

1 Complete a risk assessment.

2 Select suitable access equipment and PPE.

3 Complete a resource checklist.

4 Check that equipment meets health and safety regulations.

5 Interpret manufacturer's instructions.

6 Erect a tower scaffold to a minimum platform height above 2.5 m.

7 Dismantle the tower scaffold.

8 Follow current environmental and relevant health and safety regulations throughout.

9 Store the equipment.

CHAPTER 3

PREPARATION OF SURFACES FOR DECORATION

INTRODUCTION

Preparation of surfaces is the most important task to get right within the painting and decorating trade. No matter how skilful the decorator is, unless they have prepared thoroughly, the finished product will never be of a high standard. This could lead to not being paid and/or having to repeat the work at their own expense, costing both time and money and damaging their reputation. Different surfaces require different preparation and it pays – literally – to know how to prepare each appropriately, using the correct tools and materials for the task at hand, so that ultimately the desired effect can be achieved.

By the end of this chapter, you will have an understanding of:
- preparing timber and timber sheet products
- preparing metals
- preparing trowel finishes and plasterboard
- removing defective paint coatings and paper
- preparing previously painted surfaces.

The table below shows how the main headings in this chapter cover the learning outcomes for each qualification specification.

Chapter section	Level 1 Diploma in Painting and Decorating (6707-13) Unit 117	Level 2 Diploma in Painting and Decorating (6707-22/23) Unit 215	Level 2 Technical Certificate in Painting and Decorating (7907-20) Unit 203	Level 2 City & Guilds NVQ Diploma in Decorative Finishing and Industrial Painting Occupations (6572-20) Unit 676
1. Prepare timber and timber sheet products	1.1–1.7, 1–3.4, 4.1–4.3, 5.1–5.2, 5.6	1.1–1.7, 9.9, 11.4–11.8	1.1–1.6	4.2
2. Prepare metals	1.3, 1.5, 1.7, 5.6	3.1–3.8, 4.1–4.2, 4.5, 9.10	2.1–2.6	4.2, 5.4
3. Prepare trowel finishes and plasterboard	1.3–1.6, 5.1–5.3	5.1–5.7, 6.2–6.3, 6.5, 9.1–9.5, 11.1–11.3, 11.6–11.8	3.1–3.6	4.2, 7.6, 7.8
4. Remove defective paint coatings and paper	1.5–1.7, 3.2–3.4, 5.1–5.2	7.1–7.7, 7.9–7.10, 8.7	4.1–4.4	4.2–4.5, 5.5, 7.5–7.6
5. Prepare previously painted surfaces	1.2–1.4, 1.6–1.7, 5.1–5.2, 5.4, 5.5	7.1–7.4, 9.1–9.5, 9.7, 9.11, 11.6–11.8	5.1–5.4	3.3, 4.2, 5.5, 7.5–7.6

1 PREPARE TIMBER AND TIMBER SHEET PRODUCTS

Types of timber

Timber is the industry name given to wood which is used for all types of internal and external building work. Timber can be hardwood, softwood or timber sheet products (manufactured boards). All timber products need to be prepared before paint, stains or varnish can be applied. On sites and while doing **domestic work** you may come across bare timber which has just been fitted or previously painted wood surfaces which need preparing before any **coatings** can be applied.

On building sites and in domestic work, carpenters carry out first fixing and second fixing jobs. First fix refers to structural work which is completed before the property is plastered or covered up with another surface material (see pages 57–58).

INDUSTRY TIP

Always work as cleanly as possible and tidy up after yourself.

The range of work covered in first fix includes rafters, **studwork, floor joists** and floorboards. Second fix work refers to the finishing tasks such as skirting boards, window frames and cills, door frames, doors and **architraves**.

KEY TERMS

Domestic work: work carried out in someone's home or property.

Coatings: paints, varnishes, stains, etc., that are applied to surfaces.

Architrave: the moulded frame around doors or windows, sometimes referred to as a door frame.

HEALTH AND SAFETY

Always wear the appropriate PPE when carrying out any preparation work on bare or previously painted timbers.

IMPROVE YOUR ENGLISH

With a pen and paper, walk around your site or college workshop and see if you can identify the different types of timber used. Write down your findings and discuss them in groups.

Softwood

Softwood timbers such as pine, cedar and spruces come from **coniferous** or **evergreen trees** and are used for construction because, generally speaking, they grow very quickly. This abundance of raw material makes the wood cheaper to produce. Although they are called softwoods, they can still be quite hard and are capable of bearing weight. The table on the next page describes some common types of softwood.

KEY TERM

Coniferous or evergreen trees: cone-bearing evergreen trees which keep their leaves in winter.

Softwood	Description and use
Pine	• Pine trees are valued worldwide for their pulp and timber, and are used to make furniture, panelling, window frames, floors and roofing • Pine is mainly used for internal work, but can be used for external work if treated
Cedar	• There are many species of cedar trees grown around the world • Cedar wood is used to make anything from pencils and guitars to log cabins and fences. Some varieties are highly scented and act as an insect repellent
Spruce	• Spruce is used in general construction and for making crates and musical instruments. Its pulp is widely used to make paper • In construction it is used for first and second fix joinery work. Internal second fix items such as skirting boards, **dado rails**, architraves, doors and window frames are commonly made from spruce because it is relatively cheap and easy to work. It is nearly always painted because of the knotty and resinous nature of the timber

KEY TERMS

Dado: an area of wall immediately above the skirting board in a room, and separated from the wall filling by a timber, plaster or plastic strip secured to the wall.

Dado rail: the wooden decorative rail that separates the dado from the rest of the wall.

IMPROVE YOUR ENGLISH

In pairs, create a drawing using a pencil and sheet of paper of a domestic dwelling wall. Identify where the skirting board, **dado**, dado rail, wall filling, picture rail and **frieze** would be. When you have finished, share with others in your group and see if you can identify the different types of material that can be used for the dado rail and picture rail. Write down your findings and discuss them in small groups of three or four.

Hardwood

Hardwoods are more expensive than softwoods because they grow more slowly and so cannot be produced as fast as softwoods.

Hardwood timbers are used for interior and exterior work, especially for doors and windows and where a more decorative finish is required. Hardwoods can be painted but the coating will mask the aesthetics of the grain. Hardwood items are therefore more commonly finished with lacquer, varnish or French polish to protect and enhance their grain and be more pleasing to the eye.

KEY TERM

Frieze: a decorative band usually on a wall near the ceiling.

The table below describes some common types of hardwood.

Hardwood	Description and use
Oak	• Oak is a fairly hard, heavy and dense timber with a high crushing and bending strength. It is used for decorative effects such as internal doors and panelling but can also be used for high-quality cabinet making • Oak comes in a variety of colours from a rich honey colour to a yellowish brown and can be stained or varnished to make it darker • It can be used for exposed roof beams, skirting boards and panelling
Beech	• Beech is used in construction for doors, cabinets and log cabins and is pinkish brown in colour. It is not as decorative as some other hardwoods so is usually painted, lacquered or oiled • Beech wood also makes good firewood and is used for smoking fish and other foodstuffs
Mahogany	• Mahogany is used to make furniture, boats and musical instruments, and has a fine, even grain. It is reddish-brown in colour and sought after for its beauty and durability

Types of timber sheets

There are a number of different timber sheet products available which can be cut to size for both interior and exterior construction work. Each type of sheet timber requires specific preparation to ensure that it is not damaged.

Things to consider when working with sheet timber include:

• the composition and surface of the board
• health risks involved when preparing the surface
• the types of **abrasive** to use
• the most appropriate primer or sealer for the job
• the correct **paint system** to be used during the application.

KEY TERMS

Abrasive: the name given to sandpaper or abrasive paper which comes in a variety of grades; the higher the number the finer the grain.

Paint system: the name given to the coatings used such as primer, undercoat and gloss; this is the common paint system used for new skirting boards and architraves, etc.

The table below describes common types of timber sheet products used within the construction industry.

Timber sheet product	Description and use
Plywood	• Plywood is made from thin layers of wood, with the grain running in different directions, glued together under pressure • The laminates can be glued together with an internal or external grade of adhesive depending on the end-use requirement
Hardboard	• Hardboard is a thin board, and is usually flexible with a smooth, polished surface. It is composed of wood pulp and wood fibre or another vegetable fibre together with suitable fillers and bonding agents and is densely compacted under high pressure • Hardboards are **hygroscopic** and therefore it is advisable to paint the back and edges before fixing them in place to prevent any moisture penetrating though the hardboard, as this will contaminate it and make it unsuitable for use
Blockboard	• Blockboard is used to make doors, tables, shelves, panelling and partition walls. It is normally used for interior work because of the type of glues used • Blockboard is a wood-based sheet panel which is made up of a core of softwood strips which can be up to 28 mm wide. The strips are placed together edge to edge and sandwiched between veneers of softwood, which are then glued together under high pressure. The inner strips are generally made up of lightweight poplar wood or spruce • To achieve maximum strength, it is important to ensure that the core (i.e. the centre and therefore the strongest part of the trunk) runs lengthways • These types of timber sheet panels are produced in three layers, with one **veneer** sheet covering each side, or five layers, with two veneer sheets per side for better stability. This sheet material is seldom used now as it has been superseded by MDF

Timber sheet product	Description and use
MDF (medium density fibreboard) 	• MDF can be used to make display cabinets, wall panels and storage units, as well as for covering pipework, etc., during first fix tasks. It is normally painted and can be ready primed when delivered to sites. This sheet material is a hardboard (see page 96) made from wood fibres glued under heat and pressure • There are various reasons why MDF may be used instead of plywood: it is dense, flat and stiff, it has no knots in it and it is easily cut on a machine. It is made up of fine particles and does not have an easily recognisable surface grain, so MDF can be painted to produce a smooth, quality surface. It can also be cut, drilled, machined and filed without damage to the surface
OSB (oriented strand board) 	• OSB is a type of engineered wood, similar to blockboard, formed by adding adhesives and then compressing layers of wood strands (flakes) in specific orientations. OSB may have a rough and variegated surface with individual strips of around 2.5 cm × 15 cm (1.0 × 5.9 inches), lying unevenly across each other. It comes in a variety of types and thicknesses • OSB's properties make it very suitable for load-bearing structures in the construction industry and it is now more popular than plywood. Its most common uses are as a protective covering within walls, flooring and roof lining

HEALTH AND SAFETY

Always wear dust masks when cutting, drilling or abrading MDF because it is **carcinogenic** and therefore very dangerous to your health. MDF contains a substance called urea formaldehyde, which may be released from the material by cutting and sanding, potentially causing irritation to the eyes and lungs. Proper ventilation is required when using MDF, and face masks should be worn when sanding or cutting MDF with machinery.

KEY TERMS

Hygroscopic: tending to absorb moisture from the air.

Veneer: a thin decorative covering of fine wood applied to a coarser wood or other material.

Carcinogenic: something that can cause cancer.

Timber and timber sheet defects

There are a variety of defects that can affect new or old timber and timber sheet products. Many of these defects are caused by accidental damage or through wear and tear, but they can also occur naturally.

All defects need the correct treatment and preparation before applying coating systems. Correct and thorough preparation is the key to high standards of work being produced when finishing coatings are applied. Thorough preparation of surfaces provides the foundation for high standards of finish that not only look good but will last longer too.

The table on the next page identifies some of the defects found in timber and describes how to prepare the wood to rectify them.

KEY TERMS

Sap: a sticky, glue-like material which is sometimes known as the blood of the tree.

Exudation: oozing; the release of a liquid.

Adhere: to stick to a substance or surface.

Seal: to apply a coating such as primer or knotting solution to the surface of a material to act as a barrier and provide protection.

HEALTH AND SAFETY

If using rags to apply knotting solution or other solvent-based materials, always fully spread out the rag after use until it has dried, to avoid spontaneous combustion.

Defect	Description and treatment
Knots	• A knot is a place in the timber where a branch was formed during the growth of the tree • If knots are not sealed with a knotting solution, they produce **sap** that can bleed through the paintwork and stain the paint finish. Sap can also break the paint film in some circumstances, such as the timber not having been evenly coated with primer, etc. • **Treatment:** Dust off and seal with a knotting solution. Apply the knotting solution with a rag by wiping over the knot to seal it. This will avoid the need to clean out brushes
Resin exudation	• **Exudation** is where the natural oily resin in timber comes to the surface and stains the wood. Some softwood (c.g. Columbian pine and pitch pine) are extremely resinous and exude resin all over the surface of the wood, not just from the knots. This makes it difficult for paint to **adhere** to the surface • **Treatment:** By applying an aluminium primer, the resin will not penetrate the timber as the primer forms a metallic barrier to stop the resin from seeping through
End grain	• This is the pattern seen when wood is cut across the grain. This surface is much more absorbent than the face of the wood because the cells that used to suck up moisture into the tree have been cut across • **Treatment:** End grain must be **sealed**, usually with two coats of primer, which will help to prevent liquids being absorbed into the timber before further preparation is carried out, such as **making good** (see page 124)

Defect	Description and treatment
Cracks 	• Due to timber being a natural product, excessive heat or very dry conditions can reduce its moisture content, which can lead to splitting and cracks forming along the grain and other damage. Timber should therefore always be stored in dry conditions • **Treatment:** The area surrounding the crack or split will need to be lightly rubbed down using a fine abrasive, with the grain rather than against it. It will then need to be primed and then made good using stopper (see page 139)
Moisture content 	• Although not necessarily a defect, it is important to consider the moisture content of timber. Wood shrinks as it dries and swells when it absorbs moisture, which may affect the paint finish • If timber with a high level of moisture content is painted, it is likely that the coating will break down through a series of defects, such as blistering, wet rot and fungal growth • A moisture meter can be used to test the amount of moisture in the timber. A result of more than 20 per cent moisture content may result in **blistering** (when warm conditions cause raised bumpy areas to occur as moisture tries to escape) leading to the paint film flaking
Open joints 	• Open joints form when timbers that have been **butted** together during construction (two ends of separate timber surfaces pushed together to form one surface) shrink away from each other due to drying out, etc., causing a gap. The open joints will need to be treated in the same way as splits and shrinkage cracks • **Treatment:** The gap should be filled with the appropriate filler, then lightly rubbed down using a fine abrasive. It will then need to be primed after dusting off
Glue residue 	• When preparing timber/timber products, it is common to come across glue residue where timber has been constructed and glued together by carpenters and joiners • **Treatment:** The dried glue residue can be removed using a scraper or chisel knife or, if it is still wet, by wiping it off with a rag and appropriate solvent • Always take great care not to cut yourself or others when using these sharp tools

➡

Defect	Description and treatment
Protruding nail heads 	• Protruding (sticking out) nail heads, which extend above or beyond the surrounding surface, occur when a carpenter has not punched nails fully below the surface during construction • **Treatment:** Drive the protruding nail heads below the surface with a nail punch and hammer, then the hole can be filled
Nail holes 	• Nail holes are the opposite of protruding nails – the nails have been punched in below the surface, causing holes which will need preparing before applying coatings • **Treatment:** The holes will need to be filled and then made good
Delamination 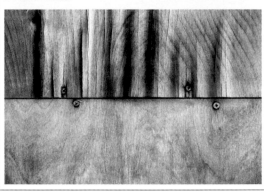	• Delamination is the separation of layers or laminations through failure of the **bond** (i.e. the layers that have been stuck together). This makes the layers separate and happens at the edge of the timber surface. When people walk on the boards, the vertical (up and down) movement between the layers starts to stretch the coating and fracture it • **Treatment:** The surface needs to be replaced

Physical properties of surfaces

- **Tactility**: Timber is not only visually attractive, it can also be very pleasant to touch, or **tactile**, even before the surface has been coated with paint or varnish. Do not run your hands over rough timber, however, or you will get splinters.
- **Porosity**: This term is often used loosely as an alternative to **absorbency**. To be precise, however, the porosity of building materials such as brick, stone, plaster and timber is the ratio of pore space to the total volume of the material. More porous materials have relatively more pore space.
- **Capillarity**: Pores do not usually absorb water to their full capacity – their rate of absorption is determined by their **capillarity**. Some surfaces are very absorbent, like blotting paper, and some surfaces, such as plastic, are non-porous and do not soak up liquid at all.
- **Adhesion**: If the surface of the timber contains small pores, it will draw the paint into the surface and help the paint to **key** on to the timber.

KEY TERMS

Tactile: relating to the sense of touch.

Absorbency: how much fluid will be absorbed (incorporated) into a material. Highly absorbent surfaces such as softwoods suck up primers and paints, impairing the finish.

Capillarity: the rate at which liquid is drawn into a material through pores or small tubes.

Key: a surface that is naturally porous or has been roughened by abrading, to help paint adhere to it.

Decontaminated: to clean a substance from a surface.

ACTIVITY

Take two small sheets of glass and hold them lightly together. Immerse the lower edge into a bowl of water and see how high the water is drawn up between the sheets of glass. Open the gap slightly to see whether the amount of water rises to a different level. This demonstrates capillary action. The bigger the gap, the less the water should rise.

Chemical properties of surfaces

Surfaces have a range of different properties.

- **Alkaline:** A substance that has a pH greater than 7. Alkalis form a caustic or corrosive solution when mixed with water. Alkalinity is the ability of water to resist changes in pH that would make the water more acidic.
- **Acidity:** This is the level of acid in substances such as water and soil. This needs to be checked before any building work is carried out so that it can be treated.
- **Inertness:** The surface has no power of action, motion or resistance. Inert is the opposite of active.
- **Soluble:** The surface or substance can be easily dissolved, especially in water or materials with a high water content. For instance, salt is soluble.

- **Salinity (salt content):** The quantity of dissolved salt content of the water. Salts are compounds and include sodium chloride (normal table salt), magnesium sulphate, potassium nitrate and sodium bicarbonate, which dissolve into ions. The concentration of dissolved chloride ions is sometimes referred to as **chlorinity**.
- **Water (or moisture) content:** The quantity of water contained in materials such as soil (called soil moisture), rock, ceramics, crops or wood.

Timber and timber sheets preparation processes

You will need to use a variety of materials and the correct processes in order to prepare surfaces without damaging them.

Cleaning agents

It is essential that all surfaces are clean and grease free to ensure a good finish. On previously painted surfaces warm water should be used with either sugar soap, washing soda or detergent (a small squeeze of washing-up liquid is a good option, because it is readily available). Stubborn marks may require a dilution of household bleach in cold water. After washing down with the cleaning agent, the surface will need to be **decontaminated** with clean water before decorating.

INDUSTRY TIP

Always wash down a wall surface from the bottom and then rinse from the top to avoid streaks on the surface.

Solvent wiping

Solvent wiping is the removal of residues from surfaces prior to applying coatings. It involves using white spirit, methylated spirit or acetone on a cloth to degrease and clean an area. Solvent wiping is convenient because the surface does not need to be decontaminated after use.

Solvents are also used to dissolve or disperse the paint film and make a paint mixture thin and fluid enough for easy application. The solvents/liquids mostly used by decorators are white spirit, turpentine and methylated spirits. These are all volatile spirits, so must be kept away from naked flames.

White spirit is used for thinning oil paint and varnish to the correct **consistency** and for wiping paint splashes from surrounding surfaces. It is also used for washing paint from brushes and equipment when tasks have been completed.

Turpentine is used in the making of paint and can also be used to thin paint and varnishes to the correct consistency.

Methylated spirit is used in the making of cellulose materials such as shellac knotting (see page 103) and cellulose paint and can also be used to thin them. Equipment and tools must be cleaned with this if cellulose materials have been used.

> ### KEY TERM
>
> **Consistency:** refers to what the coating is like in terms of its viscosity (thickness), which can be altered by adding thinners or solvents. In terms of paste, it refers to how thick or thin the material is in use. It is important to have the correct thickness to ensure papers do not become over-soaked.

> ### INDUSTRY TIP
>
> Remember, when diluting coatings, you can always add but not take away, so be careful not to add too much straight away.

Abrasives

Dry **abrading** refers to rubbing down surfaces during preparation; this type of preparation is one of the main tasks completed to produce a decent finish. There are several types of **abrasive papers** used when rubbing down and keying surfaces. It is important to use the right type and correct grade of abrasive paper when preparing treated and untreated timbers to avoid causing damage.

▲ Figure 3.1 Abrasive paper for use with a drum sander

Types of abrasives used in preparation

Using an abrasive that is too coarse can damage the **substrate** and leave scratches that may show through when painted. On the other hand, using an abrasive that is too fine may be ineffective at removing or levelling surface imperfections. Badly prepared substrates will require making good before repainting.

> ### KEY TERMS
>
> **Abrade:** to remove part of the surface of something by rubbing down with an abrasive material such as silicon carbide paper (commonly known as 'wet and dry').
>
> **Abrasive papers:** used when rubbing down to make another surface become smooth. Such materials are very important to a decorator during preparation, as inadequate rubbing down of a surface will result in a poor paint finish.
>
> **Substrate:** surface to be painted or decorated.

The table below shows some types of materials used for abrading timber, along with their properties, characteristics and uses.

Abrasive type	Material/backing	Glue to hold grit	Characteristics	Use
Aluminium oxide	Bauxite/grit with either a paper backing or cloth backing	Soluble bone glue	• Comes in disc, belt or sheet form, and may be used by hand • Extremely hard-wearing, long lasting and very economical • Grades available include 60, 80, 100, 120, 150, 180, 210, 240 and 270, and range from coarse/rough to fine/smooth	• Dry
Garnet paper	Natural semi-precious mineral on either a paper or cloth backing	Soluble and non-soluble glues	• Harder wearing than glass paper	• Paper backing for dry abrading • Cloth backing for wet abrading
Glass paper	Made from ground-up glass particles, with either a paper or cloth backing	Bone glue	• Used on dry timber and painted surfaces • Various grades available, ranging from 40–60 (rough or coarse) to 240 and upwards (very fine)	• Dry

Knotting

Knotting is the term given to shellac/white/patent knotting solutions. These materials are used as sealers to prevent resin, marks and stains bleeding through the final paint surface. Knotting solution is, as the name suggests, most often used to seal knots in timber, but it can also be used on other stains such as felt-tip pen marks and tar splashes.

Be very careful when using knotting solution as it is highly flammable. It is best kept in a knotting bottle or a glass container which prevents the knotting solution evaporating and drying out.

Applying knotting solution

- All the surfaces should be clean and dry before applying.
- Knotting solution is best used with a brush, but a clean cloth can also be used.
- Take care to cover the knot or stain entirely to fully seal it; the knotting solution should dry quickly.
- When it is completely dry, the surface coating system can be applied. If a brush has been used to apply the knotting solution, it should be cleaned with a cleaning solvent (see the manufacturer's instructions on the bottle).
- If a cloth has been used, make sure that it is opened up and left to fully dry before disposing of it appropriately.

▲ Figure 3.2 Knotting bottle and brush

▲ Figure 3.3 Knotting solution

▲ Figure 3.4 Wood stopping

Priming

Priming is the first stage of applying a coating system to a surface and is the foundation of the entire paint system. A coat of primer provides a durable and protective coating that acts as a bridge between the substrate and the rest of the paint system. With **porous** materials such as softwoods, some of the primer will be absorbed into the wood, but it will leave enough binding medium on the surface for the subsequent layers of paint to adhere to. Although it can be applied with a brush, roller or spray gun, a better finish is achieved by applying the primer with a brush, even if further layers of paint will be applied with a roller or spray. See pages 105–106 for further details of priming.

KEY TERM

Porous: a solid material having small spaces or voids that enable it to absorb liquids.

Stopping

Stopping is the preparation process of filling small cracks and holes; stopper has a stiff paste consistency. It should be used for exterior work as it is more stable in damp conditions. The stopper is pressed into holes and cracks with a small filling knife and then levelled off.

Filling

Filling refers to the application of powdered fillers to cracks, holes and indentations on surfaces. Fillers come in single-pack and two-pack varieties. There are also plastic wood fillers available, which are used to fill holes and indentations on timber surfaces that are to be varnished or stained. Plastic filler is available in a range of colours to match the surface it is to be applied to.

Single-pack fillers

The most common fillers are the single-pack powder fillers used in the painting and decorating trade. These are good all-round fillers, but they are prone to shrinkage, so be prepared to fill larger holes and cracks more than once to get them level.

To prepare this type of filler, the powder is mixed with clean cold water until it becomes a smooth consistency. When mixed, the filler is usually workable for 30–40 minutes and sets hard within a couple of hours. When dry it can be sanded back to a smooth finish ready for decorating.

INDUSTRY TIP

Single-pack fillers can be used on timber, masonry and plasterboard surfaces.

Two-pack fillers

These fillers are often referred to as two-part or deep-hole fillers. These types of filler **cure** by a chemical reaction. When the two parts of the filler are mixed together, the mixture sets within a few minutes and is ready to be rubbed down within about 30 minutes. This type of filler can be used for larger repairs, as it doesn't shrink or crack, but it is harder to rub down (abrade) than powder filler. It is ideal for repairing rotten window and door frames, and is very tough – it can be drilled, screwed into and even planed. It can also be used for interior and exterior repairs but is more expensive than single-pack filler or stopping.

▲ Figure 3.5 Two-pack filler

Plastic wood filler

This is a fast-drying filler/stopper for small nail holes and cracks, which can be used on exterior and interior softwoods and hardwoods. It can be rubbed down/ abraded when dry and it sticks well to bare wood. It can also be coated with wood stain, paints and varnish if using the natural version.

The application process is as follows:
1 Select the colour of the filler to match the colour of the timber or stained timber that is to be filled.
2 Use a filling knife or putty knife and fill slightly **proud**.
3 Allow to dry and then rub down with the grain until the filler is flush with the surface using a fine abrasive (silicon carbide paper).

KEY TERMS

Cure: to harden.

Proud: where the filler is applied so it is slightly higher than the surface; once dry, it is rubbed down to create a smooth, flat surface.

Types of primers used on timber and timber sheets

The table below gives information on priming wood, timber and timber sheet surfaces.

Primer	Description and method of application	Uses
Water-borne primer	• Water-based paint **thinned** (diluted) with water (to make it easier to apply). It is usually available in white • **Method of application**: Brush, roller or spray • Paint brushes and equipment are cleaned with water	• Can be applied to both interior and exterior surfaces including woodwork (softwood), old and new plaster, cement, concrete, hardboard and building boards. Can also be used as a matt finish for interior ceilings and walls • Not to be used on metal as acrylic primer contains water and may cause metal to rust
Solvent-borne primer	• Oil-based paint thinned with solvent. It is usually available in white • **Method of application**: Brush, roller or spray • Paint brushes and equipment are cleaned with white spirit or brush cleaner	• Can be applied to both interior and exterior surfaces such as woodwork (softwood), old and new plaster, cement, concrete, hardboard, building boards and can be used on metals
White and pink wood primer	• A general-purpose wood primer, available in white and pink • Provides good adhesion for undercoat and is non-toxic, unlike lead-based primers. (Lead-based primers are no longer available, but they may still be found on some old paintwork) • **Method of application**: Brush • Paint brushes and equipment should be cleaned with solvent (white spirit)	• This primer is harder wearing than water-based paint, so it can be used on exteriors • It can also be used on interior surfaces and is safe to use where children and pets may come into contact with the surface; it is ideal for use in hospitals, nurseries and places where food is stored

➡

Primer	Description and method of application	Uses
Aluminium wood primer	• A dull, metallic grey oil-based primer • **Method of application**: Brush • Paint brushes and equipment should be cleaned with solvent (white spirit)	• Wood primer is used for resinous timber (softwoods) and can be used to seal surfaces that have previously been treated with wood preservative
Universal wood/metal primers	• A solvent-based primer formulated for interior and exterior use • Based on an anti-corrosive zinc phosphate pigment and alkyd resin • **Method of application**: Brush • Paint brushes and equipment should be cleaned with solvent (white spirit)	• Used on new or bare wood, metal, plaster and masonry
Shellac knotting	• See knotting information on page 103 • **Method of application**: Brush or cloth • Brushes should be cleaned using methylated spirits	• Used for sealing knots on timber surfaces before priming

---

INDUSTRY TIPS

Aluminium wood primer can also seal old bitumen-coated surfaces.

Universal wood/metal primers are particularly suitable for multi-surface work.

Safety considerations when preparing timber and timber sheets

There are many safety considerations to be aware of when working in the construction industry. (These are covered more fully in Chapter 1 Principles of construction). This section looks at safety considerations when preparing timber and timber sheet materials.

Manual handling

Try to avoid manual handling if there is a possibility of injury. If manual handling cannot be avoided when working with timber and timber sheet materials, then you must reduce the risk of injury by following a risk assessment and safe working practices.

Timber and timber sheets come in a variety of shapes, sizes and weights, so you need to be very careful when handling them. If you cannot safely handle materials, then you should get help from a colleague or use the correct lifting and moving aids. Do not lift or carry anything that you cannot safely move.

When applying coatings to timber and timber sheets, you may need to use access equipment such as steps, ladders, podiums or towers. You must always follow instructions, risk assessments and safe working practices when working at height.

Control of Substances Hazardous to Health (COSHH) Regulations 2002

The COSHH Regulations 2002 control the use of dangerous substances such as preservatives, fuels, solvents, adhesives, cement and oil-based paint. These must be moved, stored and used safely without polluting the environment.

The Regulations also cover hazardous substances produced while working, for example wood dust produced when sanding or drilling.

Hazardous substances that may be discovered during the building process, for example lead-based paint or asbestos, are covered by separate regulations. If you do find asbestos, inform the site manager immediately.

IMPROVE YOUR ENGLISH

What control measures should be undertaken when working on older properties that may have been painted with lead-based paints? Use the HSE website and other internet sources for information to assist in answering this question.

When considering substances and materials that may be hazardous to health, an employer must do the following to comply with COSHH:

- Read and check the COSHH safety data sheet that comes with the product. It will outline any hazards associated with the product and the safety measures to be taken.
- Check with the supplier if there are any known risks to health.
- Use the trade press to find out if there is any information about this substance or material.
- Use the HSE website, or other websites, to check any known issues with the substance or material.

When assessing the risk of a potentially dangerous substance or material, it is important to consider how **operatives** could be exposed. Exposure could occur:

- by breathing in gas or mist
- by swallowing it
- by getting it into the eyes
- through the skin, either by contact or through cuts.

Waste and storage of materials

When carrying out tasks during preparation and decoration of timber and timber sheet materials, you must be aware of procedures for dealing with waste and storage of materials.

Any debris from preparation tasks should be disposed of accordingly, following all instructions and regulations. Coatings should be used up where possible and returned to their stock pots to be used for future tasks.

HEALTH AND SAFETY

If coatings are to be disposed of, then all requirements and regulations as set by industry must be followed.

Personal protective equipment (PPE)

All employees and subcontractors must work in a safe manner. This means wearing any PPE that their employer provides as well as looking after it and reporting any damage to it. When working with timber and timber sheet materials, you need to wear the correct PPE to prevent injuries. This means safety boots, safety gloves and a hi vis vest if working outside.

▲ Figure 3.6 Site safety sign

Ventilation

You must make sure that you have adequate ventilation in place when preparing surfaces. This can be as simple as opening windows and doors, but may involve using **extractors** during tasks, depending on the size and duration of the project.

KEY TERMS

Operative: another term for worker.

Extractor: a machine that removes dust and fumes from a work area.

Volatile organic compounds

Volatile organic compounds (VOCs) are materials that evaporate from everyday products such as cleaning products and paint at room temperature. VOC emissions contribute to air pollution and affect the air we breathe. The measurement of VOCs shows how much pollution a product will emit into the air when in use (the product container will have a VOC content label so that you can compare products). Solvents within coatings are the main source of VOCs in the painting and decorating industry, although the paint industry is trying to reduce emissions and low-odour paints are now available. VOC levels will vary for different paints and varnishes.

Prepare timber and timber sheet products

Checking timber and timber sheets for defects

Unless the surface to be painted is sound, coatings will not adhere sufficiently, making it impossible to cover – or if you manage to cover the surface, the coating will very soon peel and crack.

Decayed or denatured timber will need to be prepared and treated to make the surface ready for coating.

Decayed timber

When timber has decayed due to wood rot, it should be cut out completely and replaced with sound timber, with the inserted timber joints made well beyond the edge of where the decay appears to end.

INDUSTRY TIP

Treat both the new insert and the remainder of the old wood with a fungicidal wash before repainting.

An area that has become infected with mould, mildew or other kind of fungal growth will need treating with a **fungicidal** wash before decorating, or else the stains will show through the final finish and the fungi

will continue to grow. You will need to follow the manufacturer's instructions for applying the wash, and wear PPE including a mask, goggles, gloves and overalls.

▲ Figure 3.7 Fungicidal wash

Denatured timber

When timber is exposed to the weather for long periods its cellular structure begins to break down, leaving the surface dull and furry. This is known as **denatured timber**, and paint is unlikely to adhere to the **friable** surface. It will need to be treated before you can proceed.

To prepare the surface for decoration:
1 Rub down the surface with a fine abrasive to remove the dead fibres.
2 Remove any dust with a dusting brush.
3 Treat the prepared surface with a wood preservative and allow it to dry.
4 Apply a coat of raw linseed oil to the surface and leave for 15–20 minutes.
5 Remove any surface oil with a cloth and leave for a week to fully dry out.
6 Repaint surface, using the same method as for softwood.

KEY TERMS

Fungicide: a substance that destroys fungi.
Denatured timber: wood that has been exposed to UV light and become grey and friable.
Friable: easily crumbled or reduced to powder.

Preparing softwood timber and timber sheets for decoration

Softwoods need to be lightly sanded using fine glass paper diagonally across the grain, and then finished off by very lightly sanding along the line of the grain. This creates a scuffing of the surface and enables the paint to adhere to the painted surface.

However, if applying a clear or coloured stain or varnish, do not sand across the grain as the sanding marks will show through and look unsightly.

▲ Figure 3.8 Sanding softwoods along the grain line

Step-by-step process for preparing new bare softwood prior to decoration

STEP 1 Remove any loose debris such as plaster, nibs (small bumps) and bits of building material using a stripping knife.

STEP 2 Protruding nails or pins should be punched below the surface using a hammer and nail punch before using the appropriate filler.

STEP 3 When the filler is dry, abrade the whole surface using a fine abrasive. Remember to rub with the grain.

STEP 4 Dust off the surface with a dusting brush.

STEP 5 Apply two coats of shellac knotting to any knots on the face of the timber, ensuring that the coats are thin and well brushed out with no edge build-up. You will need to go slightly beyond the area of the knot. Allow for drying between coats. This should take about 15–20 minutes.

The softwood timber frame/architrave shown in the images on page 109 is now ready to receive its first coat of primer (see the table on pages 105–106 for detailed descriptions of primers).

Traditional oil-based primers should be used for external work, as they are hard-wearing, while acrylic water-based coatings are preferred on internal timber surfaces, as they dry more quickly and give off less odour.

When using acrylic coatings, it is possible to re-coat the surface the same day as soon as it is dry. When using solvent-based coatings, you need to wait until the following day to make sure that the primer has formed a solid foundation before applying the next coating in the system.

> **INDUSTRY TIP**
>
> Acrylic coatings are available as primers, undercoats, eggshells and glosses.

Preparing hardwood timber and timber sheets for decoration

The cellular structure of hardwood is a lot finer than that of softwood, so the pores are not as open or absorbent. Some hardwoods have an oily or acidic nature/texture. The finer grain on some hardwoods, such as oak and ash, makes it difficult for the paint to adhere to the surface. Because of this, a primer must have special properties to enable good adhesion.

Aluminium primers that contain leaf or flake pigment are ideal, as is a type of primer called calcium plumbate.

To prepare hardwood for decoration:
1 Abrade using a fine abrasive, remembering to rub with the grain.
2 Dust off using a dusting brush.

3 If the surface feels greasy to the touch, the surface will need degreasing using white spirit.
4 Dry off the surface or allow it to dry naturally.
5 Use shellac knotting to seal any knots (apply two coats and leave to dry).
6 Apply one coat of primer.

It is essential when painting timber that all end grains are properly primed to stop moisture being absorbed into the timber. Two coats of shellac knotting should be applied to the end grain before priming. Shellac knotting may also need to be applied to knots in hardwood and resinous areas if the surface is to be painted.

Preparing timber for varnish, oil or staining

Use the following steps to prepare timber:
1 Use white spirit on a lint-free cloth to remove any resin exudation or grease.
2 Lightly rub down using fine glass paper or garnet paper.
3 Use a dusting brush to remove dust and particles.

The surface is now ready to receive a coating of varnish, oil or wood stain.

Applying primers

It is recommended that a brush is used to apply primers, as the primer can be applied with more force and will therefore penetrate the surface. This enables the surface to be protected and prepared for the following coat within the system being used. Refer to the primer table on pages 105–106 for detailed descriptions.

2 PREPARE METALS

Types of metal

Metal is used in various construction stages, from laying the structural foundations, through first fix and second fix, to decorative finishing touches. It is used because it is strong, durable and waterproof and it can also be very beautiful.

Most metals suffer **corrosion** at varying rates and must be treated. Corrosion can be caused by contact with any kind of moisture, exposure to oxygen or hydrogen and by pollution in the atmosphere.

KEY TERMS

Corrosion: the wasting away of metal when it is exposed to water, oxygen, acid, alkali or salts.

Rust: refers specifically to the corrosion of ferrous metals.

There are two types of metal used in industry ferrous and non-ferrous.

Ferrous metals contain iron and will need to be cleared of all **rust** before painting. The table below describes common ferrous metals used in construction.

Ferrous metals	Properties	Uses
Cast iron	• Cast iron is a non-toxic corroding metal	• Handrails • Bridges • Railings • Building frames • Stairs • Columns • Radiators
Wrought iron	• Wrought iron is a corroding metal available in bar form, as well as sheets and hoops • This metal can crack if heated and may be brittle when cold. The texture of wrought iron is quite rough	• Ornamental ironwork • Pipework • Handrails • Roof trusses
Mild/sheet steel	• Mild steel, also known as sheet steel, has a high carbon content • This metal is very likely to rust if not treated or coated • It is a strong metal but can bend quite easily	• Girders • Tubes • Screws • Nuts and bolts • Garage doors

Non-ferrous metals do not contain iron and are less likely to corrode than ferrous metals. Examples include zinc, copper and aluminium. The following table describes non-ferrous metals commonly used in construction.

Non-ferrous metals	Properties	Uses
Galvanised steel 	● Galvanised steel comes in tubes, sheets and flat bars ● It is extremely resistant to **corrosion**, and can withstand saltwater, moisture, rain and snow ● Galvanised steel is lightweight, fire resistant and very low maintenance as it just needs cleaning down	● Girders ● Frames ● Roofing ● Support beams ● Piping
Copper 	● Available in tubes, sheets, wires, rods and flat bars ● It is easy to bend but also easily damaged so must be stored carefully ● Copper also tarnishes quickly (loses its shine)	● Water pipes ● Electrical wiring ● Roofing
Aluminium 	● Aluminium is very lightweight and is extremely resistant to corrosion	● Window frames ● Extension ladders ● Step ladders ● Mobile towers
Lead 	● Lead is a poisonous metal, so great care is needed when it has been used ● It is very soft, heavy and very resistant to corrosion. It also discolours to grey when exposed to air ● Always wear the appropriate PPE and follow good hygiene techniques when handling lead	● Roofing ● **Flashing** (the covering of the join with the roof) around chimney stacks ● Guttering ● Roof parapets

Types of metal corrosion

Most metals suffer corrosion at varying rates and must be treated. There are many types of metal corrosion; most of them are bad for the metal but some can also be beneficial (see later in this chapter). However, if defects in metal are not identified and treated before paint is applied, the surface may continue to corrode.

The results of this happening can range from being unsightly to being dangerous. The following table gives information on the types of corrosion which affect metal and how to treat them to prevent further corrosion.

Type of corrosion	Description and preparation/treatment
Surface corrosion 	● Surface corrosion is an orange-red coating that forms on metal surfaces, which will need to be removed to stop further corrosion ● **Preparation/treatment**: Priming and painting will further protect the surface from moisture
Pitting 	● Pitting is when holes, pits and craters form in a metal surface due to severe corrosion. It may also be caused by the heavy-handed use of tools for descaling ● **Preparation/treatment**: The holes, pits and craters will have to be filled with suitable filler if the surface is to have a smooth finish
Mill scale 	● Mill scale is a blue/black film that forms on sheets of steel and wrought iron during the manufacturing process. After a while it becomes loose and flaky and provides an unstable foundation for painting ● **Preparation/treatment**: The mill scale should be completely removed using a wire brush or a rotary disc sander, rotary brush or needle descaling gun
Sacrificial coating and cathodic protection 	● If different types of metal are used together, they may corrode at different rates. Sometimes metals are combined so that the one that corrodes faster is 'sacrificed' to protect the slower-corroding metal. This is called **cathodic protection**. An example of this is the rusting of corrugated iron sheeting – the steel underneath is protected from corrosion by a **sacrificial coating** (made of zinc) ● The silver-coloured parts in the image are designed to be sacrificial to protect the rest of the painted surface from corrosion

Factors contributing to corrosion

The main factors contributing to metals suffering from corrosion are contact with any kind of moisture or being exposed to oxygen or air pollution. This can happen at varying rates and must be treated as stated above.

▲ Figure 3.9 Structural steel must be protected from rust before painting

Metal preparation processes

You may need to use a range of tools, materials and equipment to prepare metalwork.

There are two main methods of preparing metals: **descaling** and **degreasing**. Descaling means removing the rust and scale from the metal and degreasing involves using a solvent-based liquid to wash the metal down before applying coatings.

Descaling

Aluminium oxide paper is used for dry abrading the metal.

Tools

Hand tools and power tools can also be used for descaling, as described in the following table.

> **HEALTH AND SAFETY**
>
> When using electrical tools always make sure that you check the plugs, cables, casing and the on/off switch beforehand to make sure they start and stop when needed.

Tool or material	Use
Orbital sander 	• An orbital sander has a flexible platform pad onto which various types of abrasive paper are fixed • The sander moves in a small circular or orbital motion to abrade the surface • An orbital sander is comparatively light in weight and can be used for long periods without fatigue • Orbital sanders produce a finer surface than rotary disc sanders
Belt sander 	• A belt sander has a continuous belt of abrasive paper with a flat sanding action. It abrades faster than an orbital sander (or sanding by hand) and comes in different sizes • Larger heavy-duty sanders are available for sanding floors. They are used to sand metal surfaces to remove light rust

Tool or material	Use
Rotary disc sander	• A disc sander has a flexible rubber sanding head that can be fitted with a range of abrasive papers. (**Rotary** means that the head spins round in circles) • It is suitable for removing rust from curved surfaces • A rotary disc sander takes practice to control and in unskilled hands it is liable to damage the surface or leave it uneven
Rotary brush	• A fitting that can be attached to an electric drill, a rotary brush is used for preparing ferrous metal surfaces • It is ideal for removing heavy rust from surfaces
Needle descaling gun	• This consists of an outer body containing a number of hardened steel needles which are propelled forward by a spring-loaded piston. On hitting the surface, the needles bounce backwards and forwards in a continuous action • The individual needles are self-adjusting, making it ideal for uneven surfaces and working around awkward areas such as nuts and bolts
Scrapers	• A good-quality scraper will have a hardwood handle and the steel blade will be **tempered** (treated in a certain way to make it firm and strong) • Scrapers are used to remove wallpaper, loose or flaking paint and other debris or nibs from areas to ensure that the surface is ready for painting or decorating
Wire brush	• A wire brush can remove loose rust and corrosion from metalwork • The metal teeth of the brush can be rubbed against the rust or loose flaking metal until it is fully removed

Tool or material	Use
Phosphor bronze brush	• This brush contains 100 per cent density of crimped phosphor bronze bristles (which do not rust) for an effective non-scarring action • When used, all the wire elements move as a single unit, promoting safer, scar-free cleaning on the most vulnerable stone facings
Steel wool	• Steel wool, also known as wire wool or wire sponge, is a bundle of strands of very fine soft steel filaments • It is used as an abrasive in finishing and repair work, for polishing wood or metal objects and for cleaning household cookware • Steel wool is available in seven grades, ranging from ultra-fine (0000) through to fine-to-medium (00, 0, 1) and medium-to-coarse (2, 3, 4). The coarse grades can be used to remove rust
Emery paper	• Emery paper is a type of abrasive paper used to abrade and polish metal. It has a rough-textured surface with a smooth paper backing • It is made by gluing naturally occurring abrasive mineral particles to sheets of paper with special adhesive. It can be glued to cloth to make emery cloth, or cardboard to make emery boards, and is also sold in discs to be used with sanding tools
Aluminium oxide sandpaper	• Aluminium oxide sandpaper is a paper which has an abrasive surface of grit • It is used for rubbing down surfaces to remove defects and to prepare them ready for coatings • The grit comes in various thicknesses and the number on the reverse of the paper denotes the grade; the fine paper/grit has a higher number such as 240 and the coarser paper/grit has a lower number such as 60

Degreasing

There are some specific materials that you will need to prepare metalwork to ensure that it is ready to receive finishing coatings.

Degreasing agents

Degreasing agents such as white spirit, methylated spirits and acetone are used to remove deposits from metals.

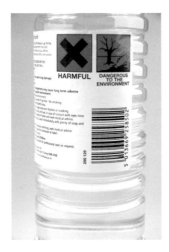

▲ Figure 3.10 A degreaser

> ### HEALTH AND SAFETY
> Be very careful when using degreasing agents, always wear the appropriate PPE and make sure that the area you are working in is well ventilated.

Mordant solution

This is used to chemically etch and prepare the surface of new, bright, galvanised metal to provide adhesion for subsequent coating systems. Surfaces need to be perfectly clean and degreased, and then rinsed with clean water prior to the mordant solution being applied. It is best to apply the mordant solution with a brush and then leave to dry. Once dried, the surface must be rinsed again and abraded before a second coat of mordant solution is applied. Once that is dried, the appropriate primer can be applied.

Rust removers

There are a variety of rust removal products on the market that will strip corrosion down to the bare metal. Before using these products, always read the manufacturer's instructions.

> ### HEALTH AND SAFETY
> When using liquid gel rust remover, remember to always wear the correct PPE and make sure that your work area is well ventilated.

▲ Figure 3.11 Rust remover gel

Types of primer used on metal

The table below identifies various primers that can be used for metals and gives you information on how to use and apply them to the metal surfaces.

Primer	Description	Use and application method
Mordant solution	• A pre-treatment for galvanised metal/steel which changes the surface to ensure the **adhesion** (sticking) of various coatings. Once applied, the surface blackens	• Used to chemically etch and prepare the surface of new, bright, galvanised metal so that paint systems adhere to it • **Application method**: Applied by brush
Zinc phosphate	• A special **rust-inhibitive** primer (meaning it will stop the surface rusting) • Paint brushes and equipment must be cleaned with white spirit	• Suitable for all ferrous metal surfaces • **Application method**: Can be applied by brush or roller

Primer	Description	Use and application method
Single- and two-pack primer	• There are two basic types of primer: single-pack (a prepared coating which is premixed and ready for use) and two-pack primer (two separate tins which need mixing together before applying to surfaces) • With two-pack primer, one tin contains the colouring and the other contains the hardener. When combined, these two resins react to produce a hardened solution	• These primers are applied directly onto a cleaned steel surface to preserve the surface and prevent corrosion • They have excellent chemical and solvent resistance and can be overcoated with epoxy (see page 161), polyurethane or acrylic coatings • **Application method**: Can be applied by brush, roller or spray
Etch primer	• Designed for retreatment of clean ferrous metals to ensure adhesion of the paint system to the surface • Paint brushes and equipment should be cleaned using the cleaning agent recommended by the manufacturer	• For pre-treatment of untreated surfaces such as aluminium, galvanised iron, zinc, copper, brass, lead, tin, clean iron and steel • **Application method**: Can be applied by brush or spray
Water-borne primer	• A quick-drying water-borne primer that can be used on all metals, including galvanised steel	• Suitable for internal and external surfaces, it can be used under oil-based and water-borne paints • **Application method**: Can be applied by brush or roller
Universal primer	• This is a solvent-borne primer formulated for interior and exterior use on new or bare wood, metal, plaster and masonry	• Based on an anti-corrosive zinc phosphate pigment and alkyd resin, it is particularly suitable for multi-surface work • **Application method**: Can be applied by brush or roller

Safety considerations when preparing metal

As with the preparation of any surface, safety considerations mean you need to be aware of regulations relating to manual handling, COSHH, disposal of waste created when preparing surfaces, storage of materials and the need to use appropriate PPE.

For metal preparation specifically, you should be very careful of flying debris when scraping, chipping and removing rust. Safety goggles or face masks should be worn as well as dust masks, gloves and overalls to protect yourself.

As with other coatings, you should make sure that the area where you are applying primers to metal after preparation is ventilated to remove fumes.

VOCs (volatile organic compounds) are found in many building materials and are partially responsible for the 'new paint' smell. VOCs are unstable chemicals that release gasses that are very harmful to people and the environment, which is why their use is now regulated by government. When using primers and coatings to protect and decorate metals, always make sure that you follow guidelines.

3 PREPARE TROWEL FINISHES AND PLASTERBOARD

Surfaces for plasterboard and trowel finishes

Most of the tasks you will perform as a painter and decorator will involve applying paint systems and other wall coverings to **trowelled finishes** and plasterboard.

Trowelled finishes is the term used for wall and ceiling coverings made from plasterboards, which come in square-edged and feather-edged squares (see opposite page). Other finishes are created with **gypsum plaster** for internal work and **cement render** for external work. Brickwork and **blockwork** (see page 120) are not usually painted, but fashions change and there is a current trend for exposed brickwork on internal walls. Before applying coatings to brick and blockwork, you must make sure that the surface is sealed with stabilising solution.

This section looks at how to prepare these substrates for decorating.

Plasterboard (square-edges and feather-edged)

Plasterboard is a type of smooth building board composed of a layer of gypsum, on either side of which is a layer of very stout paper or fibre. The paper adheres tightly because the gypsum contains a small quantity of glue. The light grey or ivory side should face out, as it is pre-primed ready for decorating. It comes in **square-edged** (sometimes called untapered) and **feather-edged** (tapered) form. Tapered edge plasterboard allows space for taping and filling, which allows a perfectly **flush** (i.e. level) surface to be achieved.

▲ Figure 3.12 Plasterboard

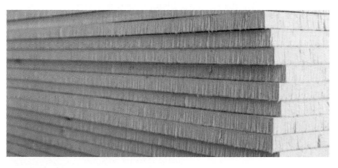

▲ Figure 3.13 Gypsum plasterboard

Plasterboard is widely used in construction as it is cheaper and quicker than applying wet plaster. The use of just plasterboards and tape is known as dry lining. The boards are attached by either sticking them with adhesive to walls (this is known as **dotting and dabbing**) or by fixing them to timber studwork with clout nails or plasterboard screws. The nails and the joints will have to be filled and taped to ensure that the surface is smooth.

Plasterboard is usually installed by a plasterer, but as a decorator it is important to be able to **patch up** small holes and cracks to make good the surface before applying finishing systems.

Care should be taken when storing and handling plasterboard because it is very easy to break the boards and cause damage to the edges and corners if not handled correctly. Do not store it directly on the floor but raise it up on timber battens to prevent damp seeping through the layers.

As plasterboard is very porous, it should be kept dry to prevent it from warping and the paper facing separating from the plaster.

IMPROVE YOUR MATHS

Measure a wall area in your classroom. Work out how many plasterboard sheets you would need to cover this area. Check local supplier catalogues for plasterboard sheet sizes.

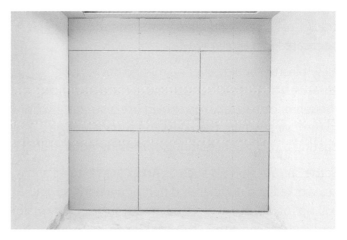

▲ Figure 3.14 Plasterboard wall

Gypsum plaster

Gypsum plaster is a material applied to internal surfaces such as ceilings and walls. It is usually smooth and tactile, although a rough surface can be achieved if required. Gypsum plaster is usually applied over plasterboard to give a better, longer-lasting finish for both paint and wallpaper. It comes as a powder and has to be mixed with water before use. Unlike mortar and cement, it remains fairly soft when it dries, so it can be sanded down and cut with metal tools. It is often moulded into decorative shapes for ceiling roses, coving and **niches**.

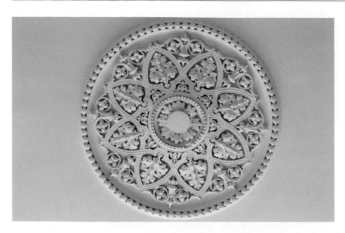

▲ Figure 3.15 A ceiling rose made of gypsum plaster

Gypsum plaster is an **inert** material, which means it does not have a chemical reaction with the paint applied to it and it receives paint systems well. As the plaster surface is porous (it has tiny holes that let air and water through), a thin layer of paint should be applied to **key into** the surface (a thick layer would lie on top of the surface and quickly break down and flake).

Like timber, plastered surfaces can be affected by cracks, nail heads and open joints, and should therefore be treated in a similar way.

New plaster should never be rubbed down using abrasives, as this will scratch the surface and show though once painted. You should use a stripping knife to remove any nibs, then dust the surface using a dusting brush before painting or **making good** (filling).

Some surface finishes such as **Artex** are textured and are worked in a number of different patterns, so you will not be able to achieve a smooth surface. If a small area is damaged it may be possible to fill the area and roughen up the surface to match the pattern, but usually the whole area will have to be re-Artexed or plastered to a smooth finish by a skilled plasterer.

KEY TERMS

Niche: a shallow recess in a wall to hold a statue, vase, etc.

Key into: to seal the surface to allow further coatings to adhere to it.

Artex: a surface coating used for interior decorating, most often for ceilings, that allows a pattern or texture to be added on application. The name is a trademark of Artex Ltd.

Cement render

Cement render is used for external wall areas and can be painted if required. The surface can be smooth or rough depending on the client's choice or surface being covered.

Blockwork

Blocks are made of concrete and are heavy, but they produce strong finished work. They are used to support heavy structures such as floors and walls. Blocks are bigger than standard bricks and therefore it is quicker to erect structures.

Thin breeze blocks and thermalite breeze blocks are lighter to use but still have the same properties as standard breeze blocks because the weight of the structure is spread evenly throughout.

They are often used where the face of the blocks will not be visible, as they are not usually as **aesthetically pleasing** as bricks.

KEY TERM

Aesthetically pleasing: pleasant to look at.

The physical properties of blockwork are very similar to brickwork, which is described below.

▲ Figure 3.16 Blockwork wall

Brickwork

A brick is a block, or a single unit of baked clay, used in construction. Typically, bricks are stacked together or laid as brickwork using mortar to hold the bricks together and make a permanent structure.

Bricks are usually produced in standard sizes in bulk quantities. With their attractive appearance and superior properties, including high compressive strength, durability, excellent fire and weather resistance and good thermal and sound insulation, bricks are widely used for building, civil engineering work and landscape design.

INDUSTRY TIP

Bricks are usually rough to the touch so are not used where a smooth, tactile finish is required.

Different styles of bricks have pores of varying sizes and will absorb water by **capillarity**. Paint adheres well to brick, as it keys into the pores, but if the bricks have absorbed a large amount of moisture the surface may crack and blister as the moisture dries out. They have an **alkaline** nature and a high soluble salt content.

KEY TERM

Alkaline: a substance that has a pH greater than 7. Alkalis form a caustic (corrosive) solution when mixed with water. Examples include lime and caustic soda.

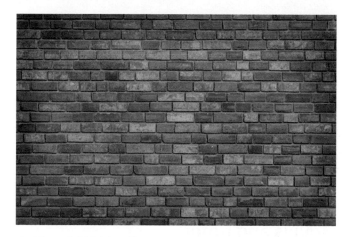

▲ Figure 3.17 Brick wall

Types of plasterboard and trowel finishes defects

Various things can influence your preparation when working with trowelled finishes and plasterboard. The age and state of repair of the structure are significant and the following defects may occur.

Cracks and shrinkage cracks

Cracks are usually found in new construction buildings and wall and surface areas where the moisture has dried out and materials have shrunk and cracked. The crack will need to be cut or **raked out** with the edge of a scraper to give the new filling something to adhere to. The area can then be wet in (see page 123) with water or coated with PVA adhesive. Once dried, the crack will need to be filled with an appropriate material (plaster, render, filler, etc.) and then rubbed down with a fine abrasive.

KEY TERM

Raked out: removal of any loose or crumbling substrates.

▲ Figure 3.18 Cracks caused by settlement of structure

Dry-out

This is a fault found in gypsum plaster, usually caused by the plaster drying too quickly in hot weather or by heat being applied to aid the drying process. It is also called **delayed expansion** and is often not visible until water-borne paint is applied. When moisture is introduced to the surface, it starts off the setting action again and can cause buckling and rippling of the plaster.

Nail heads

Masonry nails in brick or blockwork will usually be removed using a claw hammer or pinchers. If this is difficult, they need to be hammered below the surface and then the hole should be filled.

Plasterboard is attached by clout nails or plasterboard screws. These should be covered with plaster or filler to ensure that they do not show through when the finishing coats are applied.

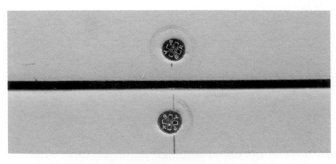

▲ Figure 3.19 Plasterboard screws correctly used during dry lining

Open joints

Open joints are where mortar (or **pointing**) between stones or brick is defective. This is a very common cause of water penetration. The making-good process is the same as for settlement and cracks.

Defective pointing

This is very similar to open joints but can be caused by poor workmanship when the bricklaying is being completed. The making-good process is the same as for open joints.

Efflorescence

This is the appearance of a fine white powder that forms on the surface of brickwork and plaster. Both materials contain soluble salts, which can originate from the raw material of bricks. However, in most cases efflorescence is caused by salts from external sources such as ground water, contaminated atmosphere, mortar ingredients and other materials in contact with the bricks. As the materials dry, the moisture evaporates and the salt is drawn to the surface, leaving a white powder on the bricks. The powder should be removed by using a stiff brush on the dry bricks.

Do not wet the surface, as the efflorescence will return when the bricks dry.

▲ Figure 3.20 Efflorescence on plaster

Saponification

Saponification is a defect that occurs when plaster or plasterboard comes into contact with lime. This contact then forms a caustic alkali which occurs when the surface is painted. It is identified as a soft, sticky paint film caused by the alkali contamination, and sometimes a brown soapy liquid exudes from the surface.

If saponification occurs over a small area it can be sealed with two coats of knotting solution. If it covers a larger area the surface should be stripped clean, washed and allowed to fully dry out before being sealed with alkali-resisting primer and then allowed to dry thoroughly before making good.

Moss and lichen

These are naturally occurring, and they grow very quickly in damp conditions. Moss and lichen are more commonly found on exterior walls, surfaces and pathways where there is excessive moisture (e.g. rain). They can be removed first by brushing or scraping and then by applying a moss and lichen remover which will inhibit their re-growth.

▲ Figure 3.21 Moss growth on external brickwork

Moulds and fungi

Mould is a furry growth of micro-organisms or fungi which grow in damp, warm conditions and must be totally removed from surfaces before painting. If you fail to totally remove and neutralise the mould/fungi, they can continue to grow under the paint, spoiling the appearance and damaging the substrate below.

Various solutions are available on the market, however, depending on the amount of growth and damaged caused. In severe cases, you may have to call in a specialist to totally remove the problem.

▲ Figure 3.22 Example of a wall with mould/fungi growth

Contamination

Contamination is a general term for anything that may adversely affect the condition of a surface or substrate, and will need rectifying or removing before you start any painting or coating system.

Friable

A friable surface is one which rubs away when you rub your hand over it. Examples include weathered cement rendering or old weathered brickwork (which is known as **spalled** brickwork). If paint is applied to a friable surface, over time it will begin to come off with the crumbling brickwork.

Chalking

This is a defect that occurs when a paint surface breaks down and becomes loose and powdery. It is caused by the weather, so usually occurs on outside walls. The powdery surface needs to be stripped down using a stripping knife and then rubbed down with abrading paper, before being primed with one or two coats of stabilising solution. Once the primer has dried, lightly rub down using a fine to medium abrasive. Make good (fill) and allow the filling to dry before rubbing down with fine abrasive to give a smooth surface ready for finishing coatings.

INDUSTRY TIP

Scrape off flaking paint with a scraping knife. If the surface below is powdery, it needs to be coated with stabilising solution prior to making good and applying coatings.

▲ Figure 3.23 Chalking on a friable paint surface

Plasterboard and trowel finishes preparation processes

Various methods are used within the industry for preparing plasterboard and trowelled finishes.

Wetting in

Wetting in the surface of deep cracks and holes before applying filler helps the filler **adhere** to the surface. You can use clean drinking water to do this or a diluted solution of PVA.

Raking out

This term refers to the removal of defective mortar or removal of flaking and loose material from rendered/plastered surfaces. Defective **mortar** between bricks needs to be removed with a heavy-duty scraper or stripping knife and then refilled with a suitable material or new mortar.

INDUSTRY TIP

Deep cracks in plaster need to be raked out to remove any loose or flaking material and then wet in prior to filling.

KEY TERM

Mortar: the material used to fix bricks together during the building of structures, etc.

Making good

The terms 'making good' or 'to make good' refer to the process of repairing or bringing something up to a finished standard or restoring it to its previous condition.

Plasterboard is very porous and will soak up any moisture, causing it to go soft or blister. You therefore need to apply a coat of sealer or primer to the surface as this will stop moisture penetrating the surface.

Any damage to the surface or substrate will first need to be repaired, as outlined earlier in this chapter.

Abrading

Various types of dry abrasives are used for making good substrates ready to receive paint systems. For surfaces such as ceilings and the top of walls you may need to use a **pole sander**.

All substrates or surfaces will either need rubbing down or keying in preparation to receive filling or coatings. The grade of abrasive that is used will depend on the condition and type of surface that needs coating. Metal can be rubbed down in any direction, but timber will need rubbing down in the direction of the grain. Plaster, brickwork and rendering should be rubbed down dry with a fine abrasive, as a coarse abrasive will damage the surface.

▲ Figure 3.24 Pole sander (the abrasive paper is attached to the head of the sanding pole)

KEY TERM

Pole sander: a hand tool that allows the user to reach higher areas when abrading surfaces.

Scraping

You will need to scrape the surface of the plaster/rendered finish if abrading or dusting off is not sufficient to remove any defects such as **bits, nibs or flaking**. To do this you will need to use a good-quality scraper/stripping knife and then dust off the surface before applying coatings.

KEY TERM

Bits, nibs or flaking: defects caused by dust and poor adhesion of paints.

Caulking and taping

Some surfaces are prone to shrinkage and movement, so cracks and holes filled with powder filler or plaster surfaces may split open. In these cases, caulking is used. Caulk, also known as **mastic** or decorator's caulk, is a flexible, waterproof filler used to fill cracks and gaps around skirting boards, doors, architraves and window frames as well as around edges of ceilings and walls.

When mastic is dry it feels like rubber and it can expand and contract with any movement or temperature change. It can be painted or wallpapered over but it cannot be abraded due to the texture of the material. You must therefore make sure there is no excess or uneven edges to the finish as it cannot be made good as with powdered fillers.

KEY TERM

Mastic: an acrylic type of caulk applied using a mastic gun (also known as a skeleton or caulking gun).

▲ Figure 3.25 Mastic/skeleton/caulking gun

When plasterboard is used on walls, there will usually be joints where the sheets of plasterboard butt together.

These will need making good so they cannot be seen. This is done by taping and then filling so that the joints do not break open.

As with new plasterwork, the surface of the plasterboard should never be rubbed down with abrasive paper as this will not only scratch the surface but will also lift the paper surface of the plasterboard sheet, severely damaging the finish.

▲ Figure 3.26 Tape and then fill the joints between plasterboards

On previously plastered or rendered surfaces you may need to fill cracks, holes or indentations to make good. The two main types of filling are as follows:

- **Proud filling**: Filler is applied so that the filling material is slightly higher than the surface. It is then rubbed down level with the surface once dried out fully. This type of filling is used on deep cracks and holes.
- **Flush filling**: The filling material is applied so it is level with the surface being filled. It is then lightly sanded down once fully dried. This type of filling is used on fine surface cracks.

KEY TERM

Flush: level with the rest of the surface.

INDUSTRY TIP

It is usually quicker to fill large cracks twice. Over-filling in the hope that the filler will shrink to fit rarely works and the area may need extensive rubbing down before refilling. So, fill, dry, rub down and fill again.

Degreasing

As with metals, the plaster or rendered surface may be contaminated with grease, oil or other contaminates like overspray from aerosols such as furniture polish or hairspray. If the surfaces have been affected by any of these contaminates then you will need to remove them by using the appropriate cleaning solution. Sugar soap and warm water is commonly used to clean the surface, but you must remember to rinse the surface down with clean water and let it dry fully prior to applying coatings, etc.

Plasterboard and trowel finishes preparation materials

The following materials are used before and during decoration tasks to prepare plasterboard and trowel finishes.

- **Plaster-based fillers**: These come ready mixed and are a convenient way of filling small holes and cracks in plaster.
- **Cement-based fillers**: These fillers are used for repairing interior and exterior masonry surfaces such as concrete, brick, block, stone and mortar. They can be applied to damp substrates and are ideal for repairing reinforced concrete. They are waterproof and come in various-sized packages.
- **Joint fillers**: These are powder-based fillers which are mixed with water until they become a fine consistency. They are used to fill plasterboard joints. Joint fillers are fast setting, so should only be mixed in small amounts to prevent them hardening before application.
- **Joint tapes**: When plasterboards have been fitted to surfaces, the joints in between each plasterboard should be covered with a fine mesh tape then a fine filler applied to prevent the joint showing through the finished paint system.
- **Reinforced corner tapes**: These are stronger tapes which offer added protection to the corners of plasterboarded surfaces.
- **Abrasives**: The correct grade to be used will depend on the surface being prepared. Lower numbers on the reverse of abrasive papers indicate a coarser grit and higher numbers indicate a finer grit.
  New plasterboard which has been taped will only need the fine filler smoothed down before applying the sealer/primer. Existing plastered or trowelled

surfaces will need a lower-grade abrasive paper if the surface has major defects such as runs, sags or drips from coatings. If the surfaces are sound, then a higher-grade abrasive paper will be needed to lightly rub down the surfaces.

Preparation materials and primers used on plasterboard and trowel finishes

There are various primers that can be used for plasterboard and trowel finishes.

Degreasing agents

See page 117 of this chapter.

Stabilising solutions

Stabilising solutions are clear, highly penetrative primers/sealers, specially formulated to bind dusty or chalky exterior masonry surfaces. They are also ideal for lowering the absorbency of very porous exterior masonry surfaces.

Size

Size is thinned adhesive that is applied to areas to seal the surface before hanging wallpaper. This product is applied with a large brush to new plaster to prepare it for paperhanging. It contains an adhesive which helps the paper slide over the surface for pattern matching.

Fungicidal wash

An area that has become infected with mould, mildew or other kind of fungal growth will need treating with a fungicidal wash before decorating. Failure to do so will mean the stains will show through the final finish and the fungi will continue to grow. The manufacturer's instructions must be followed when applying the wash, and PPE including a mask, goggles, gloves and overalls should be worn.

▲ Figure 3.27 Glue size

Stain block

This is a water-borne primer for use on interior walls and ceilings. It stops the migration of stains through the paint finish and can also be used to seal and bind dry, friable and powdery surfaces such as plaster, limestone and distemper. It is useful for covering up hard-to-conceal stains such as ink, scuffs and fire and water damage. Before use the surface must be cleaned of defective or poorly adhering materials, efflorescence, dirt, grease, wax and so on.

Once applied with a brush or small roller, it should be allowed to dry before further paint finishes are applied.

Primers

The table on the opposite page shows the water-borne and solvent-borne priming materials you will need for plasterboard and trowelled finishes.

Primer	Description	Use and application method
Emulsion (water-borne paint)	• Emulsion can be thinned out with water to form what is often called a mist-coat. Thin to manufacturers instructions but would typically be 20% added clean water	• Used on newly plastered surfaces, the mist-coat soaks into the porous surface to provide a key for further paint coatings • **Application method**: Best mixed up in a bucket and applied by brush to penetrate the surface. Rollers and spraying equipment can also be used but due to the consistency it can become quite messy
Stabilising solution	• Available in clear or coloured form. The clear solution has better penetrating properties and is easier to apply • Paint brushes and equipment must be cleaned with solvent (white spirit) or with water if water-based	• To stabilise old powdery surfaces before painting. Also used to seal plasterboard before paperhanging • **Application method**: Can be applied by both brush and roller
Alkali-resisting primer	• Primer designed for surfaces that are alkaline in nature • Paint brushes and equipment must be cleaned with solvent (white spirit)	• To prime new and old building materials which are of an alkaline nature: plaster, brickwork and concrete blockwork • **Application method**: Can be applied by brush, roller and spray
Spirit-based primer/sealer	• Shellac-based primer, sealer and stain killer	• It is perfect for use on interior surfaces and spot priming of exterior surfaces • **Application method**: Can be applied by brush, roller and spray

Safety considerations when preparing plasterboard and trowel finishes

Although health and safety instructions are common to all trades, there are some things that are particularly important to a painter and decorator, as outlined below.

- As a decorator you will come into contact with many chemicals and irritants so you must follow the COSHH Regulations at all times. You should apply barrier cream to protect your skin from contaminants that may cause infections or irritations such as dermatitis.

INDUSTRY TIP

Apply barrier cream at the start of every day.

- It is important to wear goggles, a dust mask and overalls when rubbing down, as the dust this creates can get into your eyes, be inhaled into your lungs and irritate your skin.
- Make sure that the work area is well ventilated, and damp down the floor when sweeping up.

- When working in a room that cannot be ventilated, it may be necessary to wear a fume or dust mask.
- Light cotton or latex gloves will protect your hands from cuts and abrasions when you are rubbing down and are comfortable to use.
- Rubber gauntlet gloves should be worn for heavy-duty work such as washing down walls or using liquid paint remover.
- Always follow the correct manual handling techniques as described in Chapter 1.
- Be aware of good housekeeping by always disposing of waste correctly and storing all unused materials safely away for future use.
- Remember when using sealers and primers that they give off fumes due to their VOC content, so always use ventilation and try to avoid using materials that can destroy the environment.

Prepare plasterboard and trowel finishes

As with timber and metal preparation, the aim of the decorator is to achieve a firm foundation on trowelled finishes or plasterboard, ready to receive finishing systems. Defective pointing or dry friable surfaces need to be made sound or the paint will not adhere.

The instructions below will provide you with the necessary information and principles on rectifying surface conditions and making good.

Brick and blockwork

- Check the surface for irregularities, holes, loose pointing and other surface defects.
- Fill, using a filling knife and hawk, with sand and cement mixed to a stiff paste with water.
- Check for efflorescence and moisture content with a moisture meter.
- Allow filling to dry thoroughly before coating with alkali-resistant primer, using a large brush.

▲ Figure 3.28 Defective brickwork

New plaster

- When the plaster is thoroughly dry, de-nib the surface using the blade of a scraper/stripping knife. When scraping, take care not to dig the blade in and score the surface.
- Mix up a light-coloured emulsion with water in a bucket to make a thin coating. Apply a mist-coat to the plaster surface with a large brush.
- When dry, check for cracks, holes and other defects, and fill as necessary.
- When the filler has dried, lightly rub down with a fine aluminium oxide paper and dust off.
- Coat with a mist-coat of thinned emulsion or a thin coating of alkali-resistant primer.

Existing plasterwork

- Strip wall coverings if necessary, and wash walls down to remove paste residue.

- If repainting, prepare the wall by using appropriate abrasive paper, tools, etc.
- When the surface is dry, rub down with fine aluminium oxide paper.
- Make good surface defects and allow filler to dry.
- Rub down with fine aluminium oxide paper and dust off.
- If the surface is to be papered, coat the wall with glue size.

Plasterboard

- Fill joints, nail and screw holes and other defects, and ensure that the joints are sufficiently taped.
- When the filler is dry, rub down lightly with fine aluminium oxide paper.
- Coat with alkali-resistant primer.

▲ Figure 3.29 Applying filler to plasterboard joints

4 REMOVE DEFECTIVE PAINT COATINGS AND PAPER

Removal of defective paint coatings

The table opposite identifies the defects that can occur to coatings over time or because of poor workmanship, and explains how these can be corrected.

Defect	Description and cause	Treatment
Blistering	• A defect where the paintwork lifts away from the surface in bubbles due to lack of adhesion • Main causes are moisture contained within new plasterwork that has been painted before it had time to dry out or introduced when a surface has been washed down and painted before it has had time to dry, resin exuding from knots and heat from direct sunlight on paintwork that is south facing	• To treat blistering, strip off loose paintwork until a solid edge is achieved. Use a scraper or stripping knife, then spot prime any bare surfaces • If the surface is badly affected, it may need to be completely stripped
Cracking or crazing	• Occurs in paintwork and is usually due to the application of a hard-drying coating (oil paint) over a softer coat (paint that has not fully dried). The top coating is unable to cope with the expansion and contraction of the previous coat as it dries • Another cause of cracking or crazing of paintwork is when paste or glue size is allowed to stray onto the painted surface and dry there • The defect occurs when coatings are applied	• If crazing or cracking has formed on the surface there is no remedy except to burn off the defective paint • Always wash off any paste or size as you go to avoid this defect occurring
Flaking and peeling	• Flaking is primarily due to poor adhesion, where the paint film lifts from the surface and breaks away in the form of brittle flakes • Often caused by moisture, it may be present in the surface before it is painted or may find an entry into the surface after painting via open joints or building defects • Flaking may also occur when the surface is loose or powdery. The presence of loose and crumbling plaster will also cause flaking	• Scrape off the flaking paint with a scraping/stripping knife • If the surface below is powdery it will need coating with stabilising solution before making good
Excessive film thickness Sags and curtains	• Excessive film thickness will result in runs, sags and curtains • These defects happen when paint has been applied too thickly or unevenly to a surface, particularly a vertical surface • Excessive film thickness takes the form of a thick line of paint like a draped curtain	• Some surfaces may only need a light rub down with fine abrasive to remove any edges before repainting • However, if the surface is very bad it will need to be rubbed down using a wet abrasive such silicon carbide (known as wet and dry) or removed by a scraper/stripping knife and then rubbed down flat

Paint removal processes

On new building projects you will probably be faced with bare substrates, but most of the decorator's work is on surfaces that have already been painted or wallpapered. These must be prepared to receive a new coating so that the finished coating will adhere with a smooth, flawless finish.

Surrounding areas must be protected during paint removal processes. The use of protective sheeting and moving items out of harm's way should take place before you begin.

The two forms of paint removal used in industry are liquid paint removers and the use of heat strippers, in the form of an electric heat gun or a liquid petroleum gas (LPG) gun.

HEALTH AND SAFETY

It is important to wear appropriate PPE, but extra precautions are required if you are removing old paintwork containing lead. Burning off lead can give off fumes that may be carcinogenic, so a respirator must be worn. The use of lead paint began to be phased out in the 1950s but the law banning its use did not come into effect until 1978, so some paintwork from the 1970s may still contain lead. If you suspect this about any old paintwork you can test it with a lead paint test kit.

▲ Figure 3.30 A lead paint test kit

Liquid paint removers

Liquid paint remover (LPR), also known as solvent-borne paint remover or paint stripper, is very efficient at removing thick layers of paint, leaving behind a smooth, glossy surface. However, because of its high VOC emissions, solvent-borne paint remover is increasingly being replaced by water-borne paint remover. The strong smell, slow drying time and the fact that brushes have to be cleaned using white spirit generally outweigh the advantages of using LPR.

Before using paint strippers, ensure that you:
- read and follow the manufacturer's instructions
- remove ironmongery from the surface to be stripped, for example door handles
- protect surrounding areas from any splashing or contamination
- protect yourself with the appropriate PPE
- make sure that the work area is well ventilated or protect yourself by wearing breathing apparatus.

When ready to open the container, be aware that the pressure that builds within the container can cause gas and liquid to spurt out when the cap is removed. This could result in damage to your eyes and skin. When opening and using LPR, avoid getting paint stripper on your skin or in your eyes, as this will cause burning and eye damage. If you do get the stripping solution on you, you should rinse it off using cold water and seek medical advice immediately.

Water-borne paint remover needs to be used with a metal paint kettle (make sure it *is* metal, as stripper can eat right through a plastic one).

Apply the stripper to the surface using an old paint brush and after a while the paint will start to blister, showing that it is ready to remove with a stripping or scraping knife. The amount of time that this takes will vary depending on the thickness of the paint.

▲ Figure 3.31 Liquid paint remover (LPR)

INDUSTRY TIP

After using liquid paint remover, it is very important to wash down the surface with clean water to decontaminate the area. Leaving chemicals on the surface can cause the breakdown of subsequent paint systems, leading to defects such as flaking or blistering.

Electric hot air stripper/heat gun

Hot air strippers or heat guns can be used on both metal and timber surfaces, particularly if a transparent coating is to be applied. They are not suitable for use on plastered and trowelled substrates.

Before using a hot air stripper, you must check that the equipment has a **PAT** certificate showing that it has been inspected and passed as safe by a competent electrician.

KEY TERM

PAT (portable appliance test): a safety test that must be carried out on electrical items to ensure they are safe to use.

Although they look similar to a hairdryer, hot air guns produce a much higher heat output and can burn, so always use them with extreme caution. The hot air is produced by an electric filament which can reach temperatures between 200°C and 600°C depending on the gun.

There is less risk of fire, cracking glass or scorching timber when using a hot air gun compared with using an LPG blowtorch.

▲ Figure 3.32 Electric hot air stripper/heat gun

The table below shows the advantages and disadvantages of using an electric hot air stripper/heat gun.

Advantages	Disadvantages
Does not scorch timber when burning off	Slower than LPG at removing paint
When used on window frames there is less chance of glass cracking	Needs power to run
Ideal for use on old or delicate surfaces and in areas of high fire risk such as the eaves of old buildings	Effectiveness can vary, e.g. when using in high winds

Follow these steps to safely remove coatings with an electric hot air stripper/heat gun:
- Hold the nozzle about 3–5 cm from the surface.
- When the paint has softened and starts to blister, use a stripping knife or shave hook to remove the paint.
- Do not hold the gun in one position for too long, as you may damage the underlying substrate.
- If possible, work from the bottom up to take advantage of the rising heat.
- Burn off in the direction of the grain.
- Direct the flame or hot air away from glass so as not to crack it.
- In some cases, two applications of heat may be necessary to remove all paint from the surface.

Water-borne paint can be difficult to remove by heat, so it is advisable to use a chemical stripper.

▲ Figure 3.33 Using a shave hook to remove the paint coating

▲ Figure 3.34 Burning off coating from a panel door

The correct sequence for burning off paint from a panelled door is as follows:

1 Make sure that the surrounding surfaces are protected.
2 Apply the heat to the mouldings on one side of the bottom panels of the door.
3 Work from the bottom to the top of the moulding removing the paint.
4 Repeat the process on the next panel.
5 Repeat the process on all the mouldings and panels up the door.
6 Apply heat to the bottom rail (horizontal panel), then the middle rail and finally the top rail.
7 Apply the heat to the **hinge stile** (the long vertical section of timber next to the hinge) from the bottom to the top.
8 Apply heat to the latch stiles and edges to remove the paint.

Removing paint using liquid petroleum gas (LPG)

LPG is generally only used to remove paint from timber surfaces. It can be used on metal, but it is less effective as it is slow and it can burn the paint surface. It should not be used on trowelled surfaces, plaster or plasterboard.

Before you begin, always check the hose and fittings of the equipment for any signs of damage. You can use a solution of detergent (washing-up liquid) to check for gas leaks. Apply the solution around all the fittings and bubbles will show if there is gas leaking. If any leaks are found, you must replace the equipment. Do not use it again as you may cause an accident.

Using LPG equipment to remove coatings:

● Before you start to burn off, ensure that a suitable fire extinguisher is nearby.
● Check whether a hot works permit is required before beginning any work.
● If you are burning off inside, remove all combustible materials such as curtains and furnishings.
● When burning off doors, start from the bottom. Heat rises and will start to soften the paintwork as you work upwards.

- Avoid burning off timber that may cause a fire, such as timber adjacent to the roof structure of a building. There are often birds' nests present or denatured timber which can easily catch fire from the flame of the torch.
- Stop burning off at least one hour before you leave the site, and always carry out a final check for smouldering timber before you leave for the day.

There are now many types of LPG guns on the market which can be used for smaller-scale jobs. Always follow the manufacturer's instructions when using these tools.

▲ Figure 3.35 LPG gun

HEALTH AND SAFETY

The hose of an LPG gun may be made from rubber and this can deteriorate over time, so check it and replace it if it is damaged.

The table below shows the advantages and disadvantages of using a LPG gun.

Advantages	Disadvantages
Low running cost	Some local authorities have banned (and many more are considering banning) the use of LPG burning-off torches because of the risks involved
Can be used when there is no electricity on site	Scorches the timber easily and should not be used on surfaces to be varnished
A very fast and efficient way of removing thick layers of paint	There is a danger that heat may crack the glass in windows
	You will need to take extra care when using a LPG burning-off torch, as there is a risk of causing a fire or burning yourself

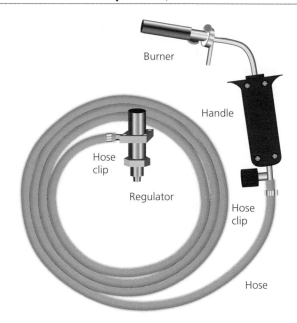

▲ Figure 3.36 Parts of a blow torch

Removal of paper

Reasons for paper removal

Removing wallcoverings is usually done for redecoration purposes. However, they may also be removed because of defects such as mould and poor adhesion.

HEALTH AND SAFETY

Take extra care when removing paper contaminated with mould, as the spores can get into your lungs. Wear goggles, a respirator, gloves and overalls and dispose of paper in a sealed heavy-duty polythene bag. Make sure your equipment and PPE are thoroughly sterilised afterwards.

If adhesive has not been applied correctly to the wallpaper or the adhesive was not of the right consistency, then the wallpaper will start to peel away from the surface and will need to be stuck back down or removed.

It is best to use a ready-mixed tub adhesive such as a border paste to re-stick wallpaper. If there is evidence of mould growth on the wallpaper, you will need to remove the wallpaper, treat the mould and then prepare the area for redecoration or wallpapering.

The type of wallpaper on the surface, the adhesive used to apply the wallpaper and the length of time the wallcovering has been on the walls will determine how easy it is to remove.

Paper removal process

When you carry out any wallpaper removal, remember to always protect the surrounding areas from being damaged or contaminated. It is best to use protective sheeting and masking tapes, after removing items from the area where possible.

Step-by-step process for removal of wallpaper using the wetting in/hand soaking method

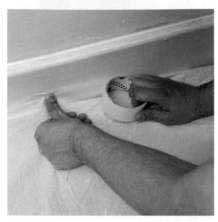

STEP 1 Cover the floor using dust sheets to protect it from any water damage, making sure that you secure the sheets using masking tape. The sheets may become very wet, so use a sign to let people know that the floor is wet.

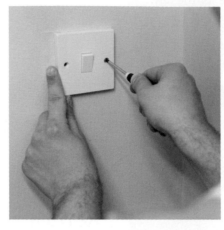

STEP 2 It is dangerous to use water where there is electricity, so turn electricity off at the mains and then loosen light switches and plug sockets so that you can strip off the paper behind them. Never switch the electricity back on until you are sure that the surfaces are dry.

STEP 3 Fill a bucket with warm water and a small amount of washing-up liquid. Now you are ready to wet the walls in, working from the bottom to the top. Apply the soapy water to the wall using a large flat paint brush or pasting brush. Starting from the bottom will help the water break the surface tension, allowing it to penetrate the wallpaper.

STEP 4 When the wall is thoroughly wet, leave the water to soak in and penetrate the wallpaper so that it softens the paste that was used to stick it to the walls. You may need to wet it a second time if it is hard to remove or use a wallpaper scourer to help with water penetration.

STEP 5 Using a scraper/stripping knife, you can now remove the paper. If the wallpaper is still hard to remove you may need to either soak the walls again or allow more time for the water to soften the paper. Remember to let the water do the work. Clear away the wallpaper as you go to remove any hazards and to maintain good housekeeping.

INDUSTRY TIP

You should persevere with wetting in until all traces of wallcovering are removed. The water needs to penetrate the paper to cause the adhesive to revert to its liquid state and release its hold on the paper.

Steam stripping

Some wallpapers/wall coverings are hard to remove by normal soaking, for example washable papers, papers that have been painted and multi-layered papers. A steam stripper will help with removal of any type of surface covering from walls and ceilings, including emulsion and other water-borne paints.

Steam strippers consist of a water tank attached to an enclosed element (similar to inside a kettle). As the water is heated it turns to steam, which travels along a flexible hose onto a perforated plate (i.e. pierced with holes). When the plate is laid against a surface, steam penetrates deeply into the wallcovering, softening the paper and the adhesive. The wallcovering can then be easily stripped from the surface.

▲ Figure 3.37 Steam stripper

INDUSTRY TIP

Never leave the steam stripper plate on a surface for long periods of time as this will blow the plaster from the wall.

Safety precautions for using a steam stripper:
- Make sure the steam stripper has a PAT certificate.
- Never let the water level drop too low.
- Never let the water dry up completely.
- Take care when re-filling as steam can scald.
- Make sure that the hose is not kinked (i.e. folded over or twisted), as this will stop the flow of the steam.
- As when stripping paper by hand, clean up as you work for safety and good housekeeping.

Dry stripping peelable papers

The term 'dry stripping' refers to removing wallpaper from surfaces without the aid of water or steam. The only type of wallpaper that can be dry stripped is vinyl wallpaper.

Vinyl wallpapers normally have a peelable top layer which should just peel off leaving a backing paper attached to the wall. However, sometimes you can encounter problems with getting hold of the peelable layer as well as the peelable layer 'running' to the edge (i.e. the strip you are peeling is getting narrower as you pull it) once you have started peeling it. Try to get a hold of the peelable layer at the top or bottom of a strip and try to get as wide a strip as possible. If you find that it is tending to run to the edge, try pulling it sideways.

Safety considerations when removing paint and paper

- Some of the surface conditions and defects that decorators must deal with can be hazardous to health if precautions are not taken.
- Buildings may have been rendered unsound by fire, flood or neglect, so carry out a risk assessment before you start work. Be aware of slipping and tripping hazards when carrying out preparation tasks.
- Remember electrical safety when working around switches and sockets. Place signs and barriers to warn other workers and members of the public of the dangers during preparation tasks.
- Take care to dispose of all debris safely and to keep your work area clean to help prevent cross-contamination.
- Check that load-bearing surfaces are safe before you walk on them and follow the Working at Height Regulations 2005 when using any scaffolding equipment such as steps or hop-ups (see Chapter 2 for more details).
- In older properties asbestos may have been used in construction, so make sure that this has been checked.
- Remember ventilation when sealing or priming surfaces after preparation and follow all VOC guidelines.

Disposal of waste

The wallpaper paste that was used to stick the paper up in the first place will be sticky again when the paper is removed, so you need to clear it up as you go to prevent trip hazards, and to stop pieces of paper sticking to your boots and being trodden onto surrounding areas. Dispose of waste in polythene refuse bags and place in suitable skips. Remember to follow all environmental procedures that apply. Some waste needs to be segregated when being disposed of to help with sustainability, recycling and reducing waste build-up.

Remove defective paint coatings and paper

When removing defective paint coatings and wallpapers from surfaces it is important to protect the area you are working on and the surrounding area. Follow the instructions given earlier in this chapter and bear in mind the following considerations:

- Using cotton or plastic dust sheets will help to protect surrounding areas.
- Removing defective paint and wallpaper from surfaces carefully and using the correct tools and equipment will help to prevent damage occurring to the surfaces.
- See pages 130–133 for the correct way to remove coatings from surfaces by heat or liquid paint stripper.

5 PREPARE PREVIOUSLY PAINTED SURFACES

Processes for preparation of previously painted surfaces

Processes for preparation of previously painted surfaces are very similar to preparing new and bare surfaces. The following information describes the processes that you will come across as a painter and decorator.

Spot prime

Spot priming is the process of applying a primer only to those regions of a substrate that need it. This technique is applicable before any interior and exterior painting on wall, metal and wood surfaces, and is often done when repainting a substrate.

Wet and dry abrading (or sanding)

As the name suggests, this preparation process uses abrasive papers either wet or dry depending on the actual finish required. One big difference is the movement used. Wet and dry abrading is done where a first-class paint finish is desired. It is used on previously treated timber to remove nibs, runs, **cissing** and other application defects to achieve a glass-like finish. This is particularly important on doors. It is not used on metal, plasterboard or plastered or trowelled surfaces.

Dry sanding requires small circles and **wet sanding** uses straight lines, alternating direction between passes. This way, each pass works to remove the scratches from the previous one.

KEY TERM

Cissing: when a coat of paint or varnish refuses to form a continuous film and leaves the surface partially exposed. The main cause of this is when paint is applied over a greasy surface.

HEALTH AND SAFETY

Put out a wet floor sign to warn people that the surface may be slippery. Wear the appropriate PPE for the task such as rubber gloves and overalls.

Step-by-step process for wet and dry abrading

STEP 1 Cover the surrounding area with waterproof dust sheets and secure them in place with either masking tape or a heavy weight. You will need a bucket with warm water and a little detergent, wet and dry abrasive and a rubber sanding block.

STEP 2 Wrap the wet and dry abrasive paper around the sanding block and dip it in the soapy water. This helps the abrasive to glide over the surface. Starting from the top, working on a small area at a time, rub down with the abrasive in a circular motion. Continue on to the next small area, dipping the block in the water frequently.

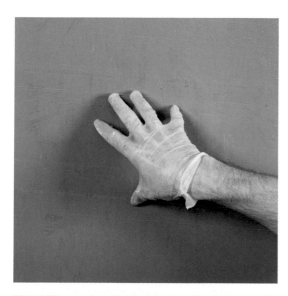

STEP 3 When you have finished abrading the whole area, check the surface for lumps, bumps and rough edges with your fingertips, repeating as necessary until the surface feels as smooth as glass.

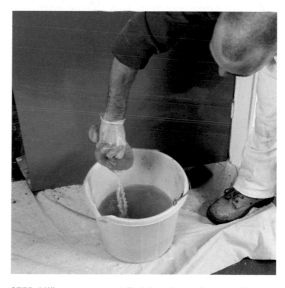

STEP 4 When you are satisfied that the surface is uniformly smooth, rinse the whole area with clean water. The surface needs to be dry before painting, so while you are waiting, clean up the work area and your tools.

Mechanical preparation

Mechanical preparation is the use of a range of different tools such as needle descaling guns and orbital sanders (see page 114) to prepare substrates such as metal and steel as well as plastered and trowelled surfaces. A range of mechanical tools was discussed earlier in this chapter.

Hand preparation

Hand preparation involves the use of hand tools and other preparation materials such as abrasive papers, dusting brushes, washing-down gear, stripping knives and scrapers to prepare and remove defects from substrates and surfaces.

▲ Figure 3.38 Caulking blade/board

Scraping

Scraping uses a stripping knife or scraper to remove a range of defects such as nibs, flaking coatings, rust and wallpaper.

Raking out

Raking out involves removing defective mortar from trowelled surfaces and defects in wood from timber surfaces.

Undercutting

Undercutting describes cutting away crumbling edges on cracks, etc., to have a clean edge to fill.

Wetting in

This term is used to mean not only wetting wallpaper before stripping it off, but also when raking out or undercutting is done. The cracks and holes should be wetted in before applying filler; this helps the filler stick to the surface. You may use a diluted solution of PVA or water to wet in.

Back filling

Where a hole or crack is too large to be filled in one go, this is a first filling which falls short of the level of the surface. The filling is allowed to dry before more filling material is added for a flush or proud finish.

Proud filling

Proud filling projects above the surface. This must be rubbed down when dry to make it level with the rest of the surface.

Flush filling

Filling that is level with the rest of the surface.

▲ Figure 3.39 Filling knife

Brushing

This term refers to cleaning down surfaces using a dust brush to remove debris and dust, etc.

▲ Figure 3.40 Dusting brush

Washing down

After stripping wallpaper, the surface must be washed down using sugar soap and warm water. Start at the bottom of the wall and move up, so the detergent used does not mark the surface of the wall, cut through the old adhesive and leave marks.

Degreasing/solvent wiping

Degreasing and solvent wiping involves ensuring surfaces are free of grease and oil before they are painted (see page 125). Washing down for a finish is similar to solvent wiping, but uses sugar soap and warm water.

Facing putty

If the facing **putty** (see below) is in good condition, simply dust it down and then apply the appropriate coating system. However, if the putty is in poor condition, it should be replaced. To replace it, totally remove the putty, taking care not to break the glass.

Stoppers used on previously painted surfaces

Putty

Putty is a cementing material made of whiting (finely powdered calcium carbonate) and boiled linseed oil. It is beaten or kneaded to the consistency of dough and is used to secure sheets of glass in sashes, stop crevices in woodwork and fill nail holes in timber surfaces.

INDUSTRY TIP

If putty is too dry and crumbles when broken, even after working it, add a small amount of raw linseed oil and continue to work the putty between your hands until it is soft. If the putty is so sticky that it does not leave your hands cleanly, add a little whiting, and then continue to work the putty between your hands until firm.

Plastic wood

Plastic wood is a fast-drying filler/stopper for small nail holes and cracks, and it can be used on exterior and interior softwoods and hardwoods. It can be sanded when dry and sticks well to bare wood. Plastic wood can also be coated with wood stain, paints and varnish.

Coloured stoppers

A stopper is a stiff paste used to fill small cracks. It should be used for exterior work, as it is more stable in damp conditions. The stopper is pressed into holes and cracks with a small filling knife and then levelled off.

Finishing plaster

Gypsum **plaster** or cement **plaster** are still used on **construction** projects, but nowadays gypsum board is used more often due to speed of installation. **Plaster** projects often are placed over a steel mesh lath that is fastened to the wall or ceiling.

Sand and cement

Sand and cement mortar is a general-purpose **mortar** which is ideal for patching, rendering, bricklaying, pointing and general repairs. It is available in a range of different sizes and can also be coloured.

Fillers used on previously painted surfaces

Powder fillers

Powder fillers are probably the most common types of filler, consisting of powder that is mixed with clean cold water. Once mixed the filler is usually workable for 30–40 minutes and sets hard within a couple of hours. When dry it can be sanded back to a smooth finish ready for decorating. Single-pack fillers are ideal for small- to medium-sized holes, scratches, cracks and imperfections, and can be drilled and screwed into once hardened. They can be used on wood, masonry, ceilings and plasterboard and are normally sold as general-purpose fillers.

Interior filler

A smooth, powdered filler for gaps and cracks. It is ideal for fixing cornices and mouldings. Interior filler is easy to sand by hand and can be screwed and drilled.

Exterior filler

Multi-purpose **fillers** can be used to fill gaps in one application. Most provide a smooth surface and can be sanded down after application if required. Ready-mix options are available for ease of use.

Ready-mixed filler

Ready-mixed micro-polymer **fillers** are very convenient and can be used for small- to medium-sized cracks and holes. However, the finished surface doesn't offer the hardness of many other **fillers**.

Fine surface filler

A ready-mixed filler that is suitable for smaller cracks in **walls** and ceilings and around doors, windows and skirtings, etc. For interior or exterior use, it can be used for repairs to brickwork, stone, plaster and wood.

Two-pack filler

Sometimes called two-part or deep-hole filler, this type of filler cures by chemical reaction. When the two parts are mixed together the filler sets within a few minutes and is ready to be rubbed down within about 30 minutes. This type of filler can be used for larger repairs as it doesn't shrink or crack, but it is harder to rub down than powder filler. It is ideal for repairing rotten window and door frames and is very tough as it can be drilled, screwed into and even planed. It can be used for interior and exterior repairs but is more expensive than single-pack filler or stopping.

Making good surfaces after removing wallpaper from a surface

Once all old wallpaper has been removed, you will need to make good the surface before applying new coatings or wallpapers.

Step-by-step process for making good

STEP 1 After stripping all the wallpaper, wash down the walls to remove any excess paste using clean warm water and a sponge. Remember to change the water regularly as clean water is needed to help remove any debris. Leaving paste on the walls may cause problems such as cissing if not totally removed.

STEP 2 You will need to wait for the walls to dry fully before making good. While you are waiting (and throughout your work), it is good practice to clear the removed wallpaper from the floor. It is important to remember to work safely. Wallpaper can be a trip hazard, and the paste is both sticky and slippery.

STEP 3 Once the walls have dried, rub them down using a medium-grade abrasive and then dust off the wall area ready for filling.

STEP 4 Mix the filler to the right consistency – it should be like whipped cream. Use a spot board (a 350 mm square of plywood) to mix on. Only mix as much as you need at a time because once mixed, the filler will start to set. Transfer the filler to a painter's hawk. Wash off any excess filler from the mixing board, ready to be used again.

STEP 5 With a filling knife, start to fill holes and cracks. Hairline cracks may need cutting open or raking out using a stripping knife, as the filler will only lie on top of the crack and will not stick to the plasterwork. Never use a filling knife to rake out, as this will damage the blade and result in a poor-quality finish.

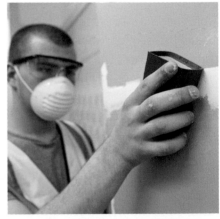

STEP 6 When the filler has dried you will need to rub down and dust off before re-decorating. Check to see if a second filling is required. If so, repeat the process.

Safety considerations when preparing previously painted surfaces

COSHH Regulations 2002

These control the use of dangerous substances, such as preservatives, fuels, solvents, adhesives, cement and oil-based paint. These have to be moved, stored and used safely without polluting the environment. The Regulations also cover hazardous substances produced while working, such as wood dust when sanding or drilling. See Chapter 1 for more on COSHH.

Volatile organic compounds (VOCs)

VOCs are vapours that evaporate from chemical substances into the air. They are found in solvents and are used in the manufacture of many coatings. They are harmful if breathed in.

Electrical safety

Only trained, competent people may work with electrical equipment. The main dangers are shock and burns (a 230 V shock can kill), electrical faults which could cause a fire or an explosion where an electrical spark has ignited a flammable gas.

Personal protective equipment (PPE)

PPE refers to all equipment (including clothing to protect against the weather) which is worn or held by an operative to protect against risks to their health and safety. Examples relevant to preparing surfaces include gloves, eye protection and high-visibility clothing. PPE for specific types of work is covered throughout this chapter.

Ventilation

Local exhaust ventilation (LEV) or extraction is used to reduce exposure to dust, fumes, vapour or gas in some workplaces. An extraction system should be easy for workers to use and enclose the process as much as possible, in order to capture and contain the harmful substance. Air should be filtered and discharged to a safe place. The system should be robust enough to withstand the process and work environment. It is important to maintain a ventilation system and tests should be carried out regularly to ensure it is working effectively.

Lead paint regulations

Working with lead can damage your health and cause headaches, stomach pains and anaemia. Other serious health effects include kidney damage, nerve and brain damage and infertility. The Control of Lead at Work Regulations 2002 (CLAW) place a duty on employers to prevent employees from being exposed to lead, or where this is not reasonably possible, to control how much employees are exposed to lead.

Disposal of waste

You are required by law to keep all areas of a construction site in good order and every place of work clean, in order to maintain a good standard of housekeeping across the site. All contractors must plan, manage and monitor their work so it is carried out safely and without risks to health. This includes careful planning on how the site will be kept tidy and housekeeping actively managed. The standard of housekeeping you achieve on site will be affected by how effective you are at material storage and waste management.

Test your knowledge

1 Which of these timbers is not a softwood?

a Oak

b Pine

c Cedar

d Spruce

2 What percentage of moisture content in timber could cause blistering and flaking to occur if painted?

a 7%

b 10%

c 15%

d 25%

3 Which of these types of surface would an alkali-resistant primer be applied to?

a Concrete flooring

b Asbestos sheeting

c Cement rendering

d All of the above

4 What causes both the fungal defects of mould growth and wet rot?

a Pollution

b Wet silicone

c Moisture

d Cold weather

5 What type of material is decorator's caulk?

a Silicone

b Flexible filler

c Powdered filler

d Dry flake

6 Which grade of abrasive paper would be needed for very rough work?

a P120

b P100

c P80

d P60

7 When abrading large, flat surface areas, which mechanical tool is best suited for the task?

a Needle descaling gun

b Wire brush

c Orbital sander

d Belt sander

8 In which form of material are liquid paint removers available?

a Gel form

b Paste form

c Powdered form

d Silicone form

9 Which of these hand tools would **not** be used to remove rust from a metal surface?

a Rotary brush

b Wire brush

c Filling knife

d Scraper

10 What is the main ingredient in knotting solution?

a Whiting

b Shellac

c Abrasive

d Dyes

Practical task

You are required to prepare a work bay for redecoration. The work bay has been papered with woodchip wallpaper that has been painted many times.

Task 1

Remove the woodchip wallpaper and prepare all surfaces, including wall area, skirting boards, door frame, door and architrave.

Task 2

Prepare a report that includes details of the tools, materials and equipment that will be needed to prepare the various surfaces and a description of the different processes involved.

CHAPTER 4

APPLY PAINTS BY BRUSH AND ROLLER

INTRODUCTION

Painters and decorators are known as 'the finishing trade' because they finish off the process when buildings have been built, refurbished or renovated. This usually happens when all the other trades have finished their work. The painting trade adds colour, decoration and protection to all surfaces that need it. No building – whether a home, factory, shop or hospital – is complete until it is decorated.

This chapter will give you an understanding of the tools, materials, equipment and techniques you will use to apply paint coatings to simple and complex areas, and how to clean, maintain and store them correctly. Chapter 3 stressed the importance of carrying out thorough preparation on all surfaces so that a high standard of finish is achieved; this chapter shows you how to achieve a high-quality finish to surfaces as that, ultimately, is what will be seen.

By the end of this chapter, you will have an understanding of:
- preparing and protecting the work area
- paint coatings and new technologies
- preparing and applying coatings by brush and roller.

The table below shows how the main headings in this chapter cover the learning outcomes for each qualification specification.

Chapter section	Level 1 Diploma in Painting and Decorating (6707-13) Unit 118	Level 2 Diploma in Painting and Decorating (6707-22/23) Unit 216	Level 2 Technical Certificate in Painting and Decorating (7907-20) Unit 204	Level 2 NVQ Diploma in Decorative Finishing and Industrial Painting Occupations (6572-20) Unit 204
1. Prepare and protect work area	1.1–1.4, 2.1–2.4	1.1–1.5	1.1–1.4	1.1–1.4, 2.1–2.3, 3.1–3.5
2. Understand paint coatings and new technologies	N/A	3.4–3.6	2.1–2.2, 3.1, 3.3	
3. Prepare and apply coatings by brush and roller	3.1–3.7, 4.1–4.5, 5.1–5.3	3.1–3.3, 3.7–3.9, 4.1–4.6, 5.1–5.2, 7.1–7.3	3.2, 3.4	4.1–4.6, 5.1–5.5, 7.3–7.4, 7.7

1 PREPARE AND PROTECT WORK AREA

When working in **domestic**, **commercial** and **industrial** areas you will need to be able to apply paint finishes to complex areas, but that does not mean that all the work you do will be complicated. All jobs will consist of straightforward tasks that you will find relatively easy, and more difficult tasks that you will need to practise to build up your work skills.

KEY TERMS

Domestic: the term used to describe people's houses and homes.

Commercial: the term used to describe shops, hospitals, office blocks, etc.

Industrial: the term used to describe factories, bridges, etc.

Environmental and health and safety regulations

As explained in Chapter 1, painting and decorating work is governed by a number of **regulations**. One of the most important of these is the Control of Substances Hazardous to Health Regulations (COSHH), as many of the materials that you will use are considered hazardous to health and need to be treated with caution. There is more on COSHH in Chapter 1.

KEY TERM

Regulations: laws and rules which have been put in place by government and must be followed.

HEALTH AND SAFETY

The Control of Substances Hazardous to Health Regulations (COSHH) 2002 explain how employers must deal with certain substances and materials so that people are not harmed by them.

When carrying out painting and decorating tasks, you need to be particularly aware of the dangers of cuts, abrasions and burns that can occur, as such injuries can become infected when working with hazardous materials. You also need to be aware that the danger of inhaling dust particles when abrading surfaces is very high, so always remember to use the correct PPE.

Another hazard that you should be aware of is that all paints and coatings contain vapours (gases) known as volatile organic compounds (VOCs). These are dangerous to people and the environment so precautions should be taken and followed (see Chapter 3 for more on VOCs).

Make sure that a risk assessment and method statement (see Chapter 1) have been carried out/produced before completing any tasks, and that warning signs are in place when working in areas where other people can be harmed.

▲ Figure 4.1 VOC label

Protecting work areas before and during work

Usually, in a domestic property which is being occupied (lived in), you will need to protect the following items prior to any painting and decorating tasks being completed:

- furniture such as sofas and other items that people sit and rest on
- wooden and tiled flooring, carpets and rugs
- door and window furniture, such as locks, handles, hinges, door knobs, letterboxes and fingerplates
- any wall-mounted fixtures and fittings such as curtains, curtain rails, lights, switches, sockets and shelves
- television and media/IT systems and lighting such as lamps and fixed lights.

It is very important to protect these items from damage as it will cost you money to replace and/or repair them. It can also cost you your reputation, so any future work you have planned could be **jeopardised**.

Before you start work, remember that anywhere you need access to will also need to be protected from damage or spillage.

KEY TERM

Jeopardised: when someone or something is put into a situation where there is a danger of loss, harm or failure.

▲ Figure 4.2 Placing dust sheets to flooring and using tape to secure them in place

▲ Figure 4.3 Removing curtains prior to decoration

As with domestic areas, items such as office furniture and lighting will need to be protected when carrying out work in commercial areas, but there may be additional considerations to take into account when working in commercial areas, such as the size of items and fixed items that cannot be moved. There is also the possibility that commercial areas may still be operating due to the nature of the business, so areas will need to be cordoned off, barriers put in place to protect the workers and signs set out to warn of wet paint, etc.

▲ Figure 4.4 Safety barrier tape

▲ Figure 4.5 Wet paint signs

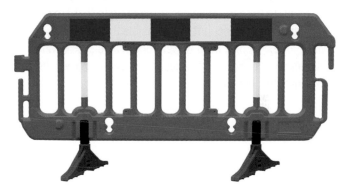

▲ Figure 4.6 Safety barrier

ACTIVITY

Speak to your supervisor or instructor and ask permission to prepare the classroom for redecoration. Remember to protect all fixtures and fittings by using the correct protective measures and also remember to use signage and barriers to prevent other people from entering the area. If you need to gain access at height, use the correct equipment and follow instructions, advice and good practice to avoid any accidents.

Workstations and machinery that cannot be moved may be present in a commercial environment, so you will need to protect them by covering them with polythene sheeting and taping up the edges to keep them in place. This will protect them from damage as well as any spillages that may occur.

Offices and shops may have equipment to keep the premises ventilated, cooled and heated. You will need to protect these items while still allowing them to be used if the commercial areas are still in use.

▲ Figure 4.7 Commercial area

INDUSTRY TIP

It is no good carefully sheeting up work areas where you are painting but forgetting about the route to that area from your van outside. If it is impossible to protect the whole area, you may need to wear disposable shoe covers. Remember to remove and replace them every time you leave and enter the work areas.

Protective materials for domestic and commercial areas

Work areas need to be protected before preparing surfaces for decorating. You should have already prepared the area when removing old paint systems or wall coverings or when repairing and making good all surfaces.

Dust sheets

A **dust sheet** is one of the most common protective materials used for domestic and commercial areas by decorators. Dust sheets protect against paint and paste splashes, as well as small particles produced when rubbing down, and small spillages. Using professional-quality dust sheets not only gives a good impression about the standard of your work but also ensures that the areas are adequately protected.

▲ Figure 4.8 Cotton twill dust sheet

Cotton twill dust sheets

The best-quality dust sheets are **cotton twill** sheets. They are generally used to protect floors and furniture and come in different sizes. They can be folded to give better protection. You can buy dust sheets in a narrow width made especially for treads and risers on staircases. They also come in different weights, from lightweight to heavy duty, and some have a protective waterproof backing.

KEY TERMS

Dust sheet: a sheet used to cover anything that needs protecting from paint or damage.

Cotton twill: a type of cotton textile that is woven with the characteristic twill pattern of diagonal parallel ribs.

Polythene dust sheets

Another type of protective sheeting used within the industry is **polythene** (a type of plastic) dust sheets. These are used in the same way as cotton dust sheets, but they are waterproof and can be thrown away after use. They can be used under cotton twill dust sheets to protect electrical equipment: the added weight of the cotton sheet helps to ensure that the plastic sheeting stays in place.

▲ Figure 4.9 Polythene dust sheet

The table on the opposite page will help you to decide whether to use a cotton twill dust sheet or polythene protective sheeting.

Type of protective sheeting	Advantages	Disadvantages
Cotton twill dust sheet	• When new or clean, they give a very professional image as they show you are careful • When used to cover the floor they will remain in place when walked on (in some places they may need taping down with masking tape) • They are available in different sizes	• They are expensive to buy and clean • They can absorb chemicals such as paint stripper • Paint spillage may soak through them • There is a risk of fire when burning off old paintwork
Polythene/plastic dust sheets	• They are inexpensive to buy • Paint spillage will not soak through them • They do not absorb chemicals such as paint stripper • They are available in different sizes	• They do not look as professional as cotton dust sheets • They do not stay in place as easily as cotton dust sheets • When wet they can become slippery • As liquid lies on the surface rather than soaking in, there is a danger of treading paint onto unprotected areas

Tarpaulin

Another protective sheeting used by the decorating trade is tarpaulin. This sheeting is made from a number of different materials including:

- PVC-coated nylon
- rubber-coated cotton
- heavy cotton canvas
- nylon scrim.

The most common size of tarpaulin used by decorators is 6 m × 4 m and because tarpaulins can protect against moisture, they are best used when washing down a surface or stripping off old wallpaper.

▲ Figure 4.10 Tarpaulin sheeting

Drop cloths

Drop cloths are another type of protective sheeting. These are made of heavy-duty cotton canvas and are hard-wearing, durable and reusable. You can also buy eco-friendly drop cloths made of recycled cotton, and waterproof versions coated with PVC or a chemical called butyl. When using a drop cloth, make sure the coated side is underneath when placing down on flooring.

Adhesive plastic sheeting

Adhesive plastic sheeting is also used to cover flooring. The sheeting comes on a roll and can be disposed of after use. The advantage of this is that it does not require additional taping to secure it.

▲ Figure 4.11 Adhesive plastic sheeting

Other types of protective sheeting

Materials such as hardboard, chipboard and blockboard can be used as protective sheeting. These types of sheeting are normally used within commercial and industrial areas as well as construction sites where there is a lot of **human traffic** nearby. Due to the hard-wearing properties of these types of protective sheeting, they can be left down until all tasks have been completed.

147

Masking tape

Masking tape is a type of protective material that is used for both domestic and commercial areas. There are a number of different types of masking tape in use by the trade, such as exterior masking tape, interior masking tape, low-tack masking tape and seven-day masking tape.

▲ Figure 4.12 Masking tape

Masking tape is a self-adhesive paper that has a variety of uses. It comes in 55 m lengths and may be 12 mm, 19 mm, 25 mm, 38 mm or 75 mm wide.

Exterior masking tape is used for exterior work such as masking up door furniture, window frames and fascia boards, and to cover surrounding areas when painting rendered, brick or pebble-dashed walls. These tapes are also waterproof.

Interior masking tape is used mainly for masking items that cannot be removed and stored, but it can also be used for taping down dust sheets to wooden floors or carpets to stop them moving and hopefully prevent tripping accidents. It can also be used to protect narrow surfaces from paint or paint remover.

A low-tack masking tape can be used for sign-writing and borders, as it does not adhere as strongly to surfaces and is less likely to pull off the underlying paint.

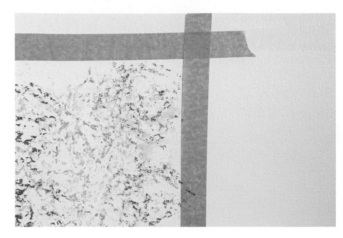

▲ Figure 4.13 Low-tack masking tape being used on a painted surface

The longer masking tape is stuck to a surface the harder it adheres, so care must be taken when removing it. Masking tape is available in different strengths of adhesion, so, as the name suggests, seven-day masking tape will be safe to leave on for seven days and will still peel off without damaging the surface.

When using masking tape to protect areas that are not to be painted, for example to form a straight line or for sign-writing, you will need to smooth it down with your fingertips to ensure the edges are sealed. If a seal is not made the paint is likely to seep through, i.e. coating will get under the masking tape and you will not have a clean finish line. When painting an area, if the coating gets onto the masking tape it can form a continuous paint film when **hard dry** (dried solid), so the masking tape will need to be removed while the paint is still tacky. Pull the tape away downwards from the leading edge to ensure a crisp line.

▲ Figure 4.14 Removing masking tape from a surface

Safety considerations when working in domestic and commercial areas

The following safety considerations should be taken into account when working in domestic and commercial areas.

- There must be safe **access** (entrance) and **egress** (exit) to the premises for the public, clients and other employees.
- In domestic or commercial environments, the property may be accessed by visitors. You may have to put up barriers to prevent people entering your work area for their own safety.
- In commercial premises, work may still be taking place in adjacent areas, so extra care will be needed to ensure that surrounding work areas are not compromised.
- The climate and weather should be taken into consideration as there may be strict regulations governing the required temperature in commercial areas. External weather conditions such as extreme cold, rain or mist may affect the progress and finish of the job.
- Internal areas may need to be cooled or heated to keep machinery and equipment working or foodstuffs at an optimum temperature.
- Any ventilation sources such as windows or air-conditioning vents should not be covered up during the protection of areas and surfaces as this could cause harm to you and others.
- Storage of materials, tools and equipment as well as the domestic and commercial items need to be considered so no damage or loss can occur.
- Fixing of the protective materials needs to be done correctly so no areas are damaged during the application of coatings.

▲ Figure 4.15 Wet floor sign

Prepare and protect work area

Follow the procedure below to ensure you take the right steps to prepare and protect both domestic and commercial work areas before the application of paint.

- Where possible, portable items should be stripped from the room/area and stored during the painting and decorating process. You may be responsible for reinstating these items, in which case you may need to label them so you know where they go.
- Remove all moveable items from the room/area and store them in other rooms, making sure that access routes are not blocked.
- Furniture or office equipment that is too large to be moved should be relocated to the centre of the room and covered with protective sheeting.

- Office equipment such as computers and printers that cannot be removed from the room must be covered carefully and unplugged if possible (check with the client first).
- If working in a kitchen, ensure that gas and electric hobs are not turned on as these could cause a fire hazard.
- If the carpet is to be lifted, roll it up with the underlay and place in the centre of the room.
- Remove all curtains, nets and blinds and carefully fold them. If possible, store them in another room to prevent any damage.
- Remove all curtain rails and fittings and store them in a safe place together ready for refitting.
- Apply masking tape to any window or door furniture that is not removed.
- Switch off the electricity supply at the mains before loosening light fittings and switches.
- Remove all ironmongery (furniture) from windows and doors to be painted. This includes latches, locks, handles, numbers, finger plates, kick plates, etc.
- Mask up (i.e. cover with masking tape) where needed and remember not to leave masking tape on longer than necessary.
- Cover all floorings with dust sheets and secure with masking tape.

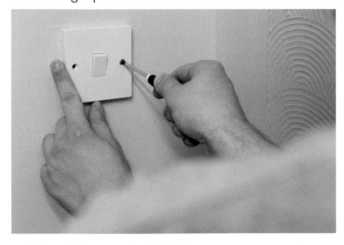

▲ Figure 4.16 Loosen electrical fittings

HEALTH AND SAFETY

Do not leave switches and sockets loose when you are away from the work area, as this could be dangerous when the mains are turned back on.

▲ Figure 4.17 Window latches and other ironmongery should be removed before painting

INDUSTRY TIP

Make sure you check for your nearest exits in case of fire when you are working in an unfamiliar building.

When working outside of a domestic or commercial property you will also need to consider the following points:

- Protect flower beds and shrubs.
- If you need to access doors or windows from the outside, make sure that you do not damage plants and ornaments. Cover with light polythene dust sheets to stop paint or debris falling on plants and pathways.
- Terraces or patios may need protecting from paint and debris as well as steps or scaffolding. Use tarpaulin to protect the ground, as it is waterproof and harder-wearing than cotton dust sheets.
- PVC guttering and pipes can be removed if required by unclipping them from the retaining clips, so that you can paint behind them.
- Make sure that there is no risk to the public and householders from paint and equipment – use warning signs to make people aware of the dangers.

IMPROVE YOUR ENGLISH

Prepare a short talk to explain to your peers the steps you need to take before painting a commercial office that is open to staff and members of the public.

② UNDERSTAND PAINT COATINGS AND NEW TECHNOLOGIES

Paint coatings and their properties

Before preparing and applying water-borne and solvent-borne coatings to surface areas you will need to do your own risk assessment, not only to make sure that you and those around you are safe, but to make sure that your finished work will be of a high standard.

IMPROVE YOUR ENGLISH

Write a risk assessment for decorating work on a corridor within your local college which is open to the public as well as other students. Discuss your findings with your tutor/supervisor.

Have you ever stopped to think why it is necessary to paint anything? When asked, many painters and decorators don't really know! Most people say that it is to make things look nice, and that answer is not wrong, but making something look nice is not *all* a painter and decorator does.

There are four reasons why a paint coating is applied to a surface. To help you remember these, you can use the acronym **DIPS**.

Each letter stands for one of the reasons for painting:

Decoration

Identification

Preservation

Sanitation.

▲ Figure 4.18 Pipes painted for identification

Decoration

We all have our own tastes – what we like and do not like – but the reason for decoration is to make something look nice. You might not like the colours that the client has chosen, but you still have to apply them to the best of your ability. With experience you might be able to advise them if you know that certain colours do not work together. There is a whole science surrounding the use of colour, with some colours said to make people feel different emotions. For example, pale blue can be calming, sunny yellows can make you happy and red can make you anxious – red is not usually used to paint hospital walls, for example. (There is more on the theory of colour in Chapter 6.)

▲ Figure 4.19 Example of colour cards used within industry

Identification

Different colours or types of surface coating can be used to identify areas or components. For example, pipework may have a British Standard (BS) colour painted on it to identify whether it is carrying gas, water or other liquids. Such industry standards ensure that all manufacturers use the same colours for identification and that mistakes are not made. See page 240 in Chapter 6 for more details about standard colours.

Preservation

Painting can stop metal **corroding** (becoming rusty) and wood from rotting, particularly when exposed to weather. An exterior door that has not been painted or that has lost its paint coating will rot and need replacing, which will cost far more than regularly maintaining the door with paint or varnish. Corroding metal can also cost lives, if it causes a structure to collapse.

Sanitation

Coating **substrates** with paint prevents germs and dirt **penetrating** (getting into) a surface, which makes the surface much easier to wash and keep clean. This is particularly important in hospitals, shops and manufacturing or where people are working with food.

> ### KEY TERM
>
> **Substrate:** a name used in industry to describe an underlying surface and surface type, such as timber, metal or plaster, onto which paint is applied.

IMPROVE YOUR ENGLISH

Make a list of the different paint colours that you see when you are in the classroom, workshop, college, local shops, office blocks and even in your own home.

Record the colours and state whether they are appropriate for the area. If you could change any of the colour schemes, what would you change them to – and why?

Types of paint

There are two main types of paint available:
- **Water-borne paint**: Where the liquid part of the paint is water.
- **Solvent-borne paint**: Where a chemical has been used instead of water to dissolve (thin out) the other components in the paint.

When paint is applied to a surface, the water or the solvent (depending on the type of paint being used) will evaporate into the air, leaving behind a solid film that forms a protective and decorative layer on the surface.

Water-borne paint consists of three main elements:
- **Thinner**: This is either the water-borne or solvent part of the paint that dissolves the other components so that it is possible to apply the paint to the surface.

- **Form-filler or binder**: This is a resin that forms the film of the paint. The binder determines how long it will last and the type of **finish** (e.g. gloss, eggshell, flat).
- **Pigment**: This gives colour to the paint and is also responsible for the paint's ability to cover the surface.

▲ Figure 4.20 Pigment is used to colour paint

Oil-based paint contains a fourth element known as the **drier**, which speeds up the drying process. Paint may also contain a **dispersant** or **dispersing agent**, which helps to keep the pigment wet, so it does not settle into a hard lump at the bottom of the can, and **emulsifiers** to keep the elements of the paint stable.

Paint dries in the following ways:
- **Evaporation**: The water or solvent turns into a vapour in the atmosphere and disappears.
- **Coalescence**: This applies to water-borne paints in which the binder is a chemical dispersed in water. When the water evaporates, the polymer particles **coalesce**, or come together, to form one thing. When all the water evaporates, the polymer particles form a uniform film.
- **Oxidation or chemical reaction**: As the liquid part of the paint evaporates, a chemical change takes place as oxygen combines with the resin and oils to form a dried paint film.

> ### KEY TERM
>
> **Coalesce:** when particles merge to form a film, particularly in water-borne coatings, they are said to coalesce. The drying process is known as coalescence.

Impact of environmental conditions on paint

There are many factors to consider when using paint, as different conditions may affect your finished work:

- **External**: Rain, snow, sleet, overcast conditions, wind, storms, sea mist, pedestrians, vehicle traffic and pollution.
- **Internal**: Dust, grease, damp, poor light conditions, occupation (i.e. is part of the area being used?), public areas.
- **Location**: Rural areas, industrial areas, coastal areas.

> ## INDUSTRY TIP
>
> Always read the manufacturer's instructions before using paint, as some paints give off vapours that can be harmful to health. Remember to wear your PPE (see Chapter 3 for more details).

All coatings rely on temperature to enable water and solvents to evaporate into the atmosphere so that the material can dry. If the weather conditions are not good, problems can occur during drying.

If it is too warm:

- the applied paint becomes too thin and does not cover the previous coating
- the paint may dry too quickly while it is being applied, due to the solvent evaporating during the oxidation process.

If it is too cold:

- the paint will not dry
- it may be difficult to apply the paint
- the paint may become too thick to apply
- condensation may form on the painted area
- surfaces may be affected by frost.

> ## HEALTH AND SAFETY
>
> Some paints give off fumes known as volatile organic compounds (VOCs) as they dry so make sure that your work area is well ventilated.

If it is too windy:

- paintwork may be covered in dust/debris, affecting the standard and quality of the completed work
- access equipment (ladders, scaffolding towers) cannot be used safely

- there may be damage to property during the application of the paint material (e.g. paint splattering on cars and flower beds below)
- it becomes unsafe to use burning-off equipment.

If it is too wet:

- the film finish may be impaired (loss of gloss, **flashing**)
- paint may not adhere (stick) to the previous coat of paint
- it may become unsafe to work
- it may delay the completion of the job.

Always make sure that the work area is well ventilated if you are working inside, not only when rubbing down or painting but also when burning off, because as the paint softens it starts to give off fumes again.

> ## KEY TERM
>
> **Flashing:** a defect that occurs in flat and eggshell finishes; it looks like glossy streaks or patches.

Types of coating

When applying coating systems to surfaces/substrates, it is important to understand the different reasons for using them and to understand the properties of coatings.

The combination of layers of paint is known as a **paint system** and may consist of many different paints. The foundation coat of paint on a new surface is the **primer** or **sealer**, and it forms a **key** between the surface and the paint. This means that the first coat bonds to the porous surface where it sinks in and grips on to it – or if the surface is non-porous, the paint film will lie on the top of it. (See Chapter 3 for more on primers.) The second coat to be applied is known as the undercoat, and this is followed by as many finishing coats as necessary.

Undercoats are designed to give a sound base for the finish, while a **finish coat** (topcoat) is the coat of paint that will be seen at the end of the job.

> ## KEY TERM
>
> **Primer:** a primer is the first coat of paint applied to a surface. The main purpose of a priming coat is to make the surface suitable to receive further coats.

Thixotropic paint

Most coatings come in a liquid form and will therefore need to be fully mixed before applying them. However, thixotropic paint is a coating that requires no mixing. It has a jelly-like structure which breaks down to a thin liquid if the paint is stirred, shaken or exposed to heat. The structure of the paint can also be broken down if over-worked by brush or roller during application.

The advantages of using thixotropic paint are that it does not require stirring before use, and it is therefore easy to apply without dripping or splashing. There is also less likelihood of **runs** and **sags** occurring, where the coating shows a long drip or too much is applied. The two main drawbacks are that the paint tends to **sweat** when stored, especially in warm conditions, and it thins rapidly if used in sunlight. This defect can often be overcome by shaking or stirring the paint and then allowing it to revert to its high viscosity.

KEY TERM

Sweat: a defect in which paint or varnish develops tackiness or thickens when it is left standing for long periods.

▲ Figure 4.21 Thixotropic paint

ACTIVITY

Take a small amount of water-borne paint and the same amount of thixotropic paint and then stir and strain the paints. Compare the two **consistencies** (i.e. thickness and texture). Check the liquids at short intervals to see if the consistency changes. Try painting a small area with each of the paints and compare their coverage and ease of use.

Paint additives

Coatings are made up of a number of substances. The information below will help you to understand the different additives, liquids and oils that can be added to coatings and the reasons why they are added.

Anti-frothing agent

This is an additive that reduces the surface tension of a solution or emulsion, and so helps to stop froth and bubbles forming and breaks up foam that has already formed. Painting with frothy paint will show in the final paint finish. Commonly used agents are insoluble oils, dimethyl polysiloxanes and other silicones, certain alcohols, stearates and glycols.

Biocides

These are added to keep bacteria from spoiling paint during storage and will also help to keep fungi and algae from growing on the applied coatings.

▲ Figure 4.22 Prevent growth of algae by adding biocides to paint before application

Water

Water is the liquid medium for water-borne coatings. These coatings can be thinned by adding water, but care must be taken to follow the manufacturer's instructions, as over-thinning will result in poor coverage.

Extenders

These are materials that have little or no **opacity** when mixed with varnish or oil. They are incorporated into paints for a variety of technical reasons, such as to prevent the settling of heavy pigments, harden the film, increase the body and slow the flow of the coating. It can be useful when painting a large area to add some extender to allow you to finish the job without altering the pigmentation of the coating.

KEY TERM

Opaque: not able to be seen through; not transparent. Opacity is the quality of being opaque.

Paint systems

When applying coatings to different surfaces and substrates, it is important to use the correct system of application. The table below gives you the correct systems that should be used to produce the right finish and to protect the substrate.

Substrate or surface	Coating systems used			
	First coat	**Second coat**	**Third coat**	**Fourth coat**
New, unpainted surfaces (to be painted with solvent-borne or water-borne coatings)	Primer/sealer	Undercoat	Gloss	N/A
	Primer	Undercoat	Eggshell	N/A
	Sealer	Emulsion	Emulsion	N/A
	Emulsion	Emulsion	N/A	N/A
	Special primer	Undercoat	Finishing coat	Finishing coat
	Stain	Stain	N/A	N/A
	Sealer	Varnish	Varnish	N/A
Previously coated surfaces (to be painted with solvent-borne or water-borne coatings)	Undercoat	Gloss	N/A	N/A
	Undercoat	Eggshell	Eggshell	
	Emulsion	Emulsion	N/A (unless covering a darker colour with a lighter colour)	
	Acrylic undercoat	Acrylic gloss	N/A	
	Stain	Stain	N/A (unless covering a darker colour with a lighter colour)	
	Varnish	Varnish		

The drying process

Coatings dry at different stages. The drying process is where a film of paint/coating changes from a liquid to a solid. There are three main ways in which paints and coatings dry:

- **By evaporation of the solvent**: As the coating dries, the solvent evaporates into the atmosphere.
- **By oxidation of the oil content**: This is a chemical process in which substances take up or combine with oxygen.

- **By polymerisation**: This is a change in the structure of a medium, in which molecules fuse together to form a solid film. The coating does not rely on an oxygen supply, as drying is caused by chemical reaction.

Coatings may dry by just one of these methods, or more likely by a combination of all three. Traditional oil paint coatings, for example, dry partly by the evaporation of the solvent and partly by the oxidation and polymerisation of the oil, while cellulose/solvent-borne coatings dry entirely by the evaporation of the solvent.

▲ Figure 4.23 Drying paint

▲ Figure 4.24 Dry paint film gauge

Film thickness

Drying time is the length of time between the application of a coat of paint and the point at which it achieves a certain degree of hardness. The following terms are used to indicate the degree of hardness achieved by the coating film.

- **Flow (dry dust free)**: The stage of drying when particles of dust that settle on the surface do not stick to the coating film.
- **Set (dry tack free)**: The stage of drying when the coating no longer feels sticky or tacky when lightly touched.
- **Touch dry (dry to handle)**: The stage of drying when a coating film has hardened sufficiently so the object or surface that has been coated may be used without marring it.
- **Hard dry (dry to re-coat)**: The stage of drying when the next coat can be applied.
- **Thorough dry (dry to sand)**: The stage of drying when a coating film can be sanded without the sandpaper sticking or clogging.

The film thickness will need to be checked against the specification to ensure that sufficient coating has been applied. The film can be checked either wet or dry depending on the circumstances of the project.

Dry paint film thickness

Figure 4.24 shows the dry film thickness being measured on a steel bridge using a dry paint film gauge. The specification, particularly in relation to industrial structures, will need to be a specific thickness when dry and will determine whether extra coats need to be applied.

Wet paint film thickness

To measure the wet film thickness during application, a **wet film comb** (also known as a **wet film gauge**) is often used. This is a very simple and quick tool for measuring the film thickness of wet paint. A wet film comb is a rectangular piece of metal or plastic with teeth of various lengths, each representing a different film thickness. The longest teeth are at the top and bottom of each side of the rectangle.

▲ Figure 4.25 Wet film comb

New technologies used in painting and decorating

A range of new technologies have been developed for use in the painting and decorating industry, as outlined below.

Nanotechnology coatings

Nanotechnology coatings or nanopaints contain **nano** structures that build a consistent network of molecules on the surface of the coating to make it super-**hydrophobic** or super-**hydrophilic**. The term 'nano' refers to coatings with a thickness below 100 **nm**.

KEY TERMS

Hydrophobic: describes a coating or material whose surface repels water. Droplets hitting a super-hydrophobic coating can fully rebound.

Hydrophilic: describes a coating or material that attracts water and tends to absorb it.

nm: nanometre – a unit of measurement used to measure the size of nanoparticles, i.e. how thick nano coatings are, and other very tiny dimensions, such as atoms. A nanometre is one billionth of a metre.

Self-cleaning coatings

Traditional coatings often lack the ability to repel dirt and other **contaminants**, leading to the dirt and dullness that can be seen on most coated surfaces. **Self-cleaning paint** is a new coating that has been developed which is both hydrophobic and oleophobic (repels oil). Self-cleaning paint is already available for masonry coatings and it is hoped that in future it may be used on cars too.

KEY TERM

Contaminants: any airborne chemical compounds that affect surfaces, such as adhesives, industrial fallout, rail dust, acid rain, bird droppings, road tar, grime, tree sap, bugs.

Anti-scratch coatings

Clear polyurethane coatings have been developed to provide a protective coating for painted furniture. This coating adheres to almost anything and, after it cures, has a very hard finish which makes it resistant to scratches.

Environmentally friendly coatings

The coatings industry is always striving to produce eco-friendly products due to the environmental impact of some solvent-borne coatings and as both consumers and regulatory restrictions demand healthier and cleaner coatings. Environmentally friendly coatings are now being produced that are odourless, less polluting and produce low to zero volatile organic compounds (VOCs).

Water-borne coatings are eco-friendly and greener because water is used to dilute them and spread out the resins within the coatings instead of organic solvents that produce VOCs.

New coatings are added to the market all the time, including recently **multi-purpose** and **renovation** coatings that can be applied to a wide range of surfaces. In some cases, minimal preparation is required, and the primer coat may not be needed.

Multi-purpose coatings

A paint coating type that can be applied to a range of different surfaces such as wood, metal, plaster, plastic, ceramics, etc.

Renovation coatings

A type of paint coating that is applied to a variety of surfaces to freshen up or change appearance, such as painting kitchen cabinet doors, tiles, etc.

3 PREPARE AND APPLY COATINGS BY BRUSH AND ROLLER

Types of water and solvent-borne coatings

When applying coatings to surfaces you will come across many different types of water- and solvent-borne coatings for both interior and exterior areas. Some coatings will be pigmented and others non-pigmented. Coatings are materials containing tiny pigment particles suspended in a binder, such as oil. Some pigments are **transparent** or **translucent** and others are opaque. Pigment particles absorb certain wavelengths of light, creating coloured paints, or they may reflect, refract and scatter all wavelengths of light, appearing white to our eyes.

> ### KEY TERMS
>
> **Transparent:** easily seen through, like clear glass.
>
> **Translucent:** allows light to pass through but things cannot be seen clearly.

Coatings is the term used within the industry to identify paints, stains, preservatives and varnishes. Coatings come in a range of different finishes such as matt, gloss, mid-sheen, silk and eggshell.

There are many types of finishing coats available and choosing the best one for the job often comes down to personal taste.

The table below will give you the information needed to choose the correct type of coating for the different surfaces that you will come across in your career as a painter and decorator.

> ### INDUSTRY TIP
>
> Often an appropriate preparatory coating may need to be considered and applied before a coating. The use of the coatings listed in the table as well as primers and preparatory coatings are covered in detail in Chapter 3.

Coating type	Description	Application and cleaning	Uses
Matt emulsion (interior)	• A water-thinned paint suitable for painting ceilings and walls. It is easier to apply than oil-based/solvent-borne coatings • Matt emulsion dries to a **flat/matt** finish	• Brush, roller or spray • Paint brushes and equipment to be cleaned with water	• Mainly used for walls and ceilings – suitable for use over plaster, plasterboard, hardboard, brickwork, cement, rendering and wallpaper
Vinyl silk emulsion (interior)	• Similar to matt emulsion paint, but with less opacity, and dries to a **sheen** finish	• Brush, roller or spray • Paint brushes and equipment to be cleaned with water	• The same as matt emulsion, but dries to a **sheen** finish which can be easily wiped down and is harder-wearing, making it more suitable for bathrooms, kitchens, hospitals and schools
Oil-based undercoat (exterior)	• A heavily pigmented oil-based/solvent-borne coating that dries to a matt finish and comes in a variety of different colours • It has good adhesion to the primer and good **opacity**	• Brush, roller or spray • Paint brushes and equipment to be cleaned with solvent (white spirit)	• Used over previously painted surfaces, timber, plaster, concrete and metalwork • It gives body and colour to a paint system and can be used over all primed surfaces, both inside and outside

➡

Coating type	Description	Application and cleaning	Uses
Gloss finish (exterior)	• Interior and exterior decorative paint used as the main protective coating in the decorating trade. It dries to a very high-gloss finish • Excellent flow when laying off (see page 258) • Very good flexibility, allowing the paint to expand and contract when dry • Good weather resistance	• Brush, roller or spray • Paint brushes and equipment to be cleaned with solvent (white spirit)	• A decorative finish for interior and exterior surfaces • Can be used on all woodwork, plaster and metalwork
Eggshell/semi-gloss finishes (interior)	• An interior decorative paint that dries with a **sheen**, also known as a **silk** or **satin** finish • As this is a solvent-borne paint it will dry to a harder finish than vinyl silk paint	• Brush, roller or spray • Paint brushes and equipment to be cleaned with solvent (white spirit)	• Decorative finish for interior surfaces including ceilings, walls, softwoods, hardwoods and metal surfaces
Masonry paint (exterior)	• A durable paint used for exterior walls (not timber surfaces) that has good opacity and is also alkali resistant • The finish is not only tough, but durable and flexible	• Brush, roller or spray • Paint brushes and equipment to be cleaned with water	• Used to protect surfaces against the weather while also giving a good decorative finish • Used on new and old cement rendering, concrete, brickwork, pebbledash and other types of masonry
Low-odour eggshell (interior and exterior)	• A water-borne coating for interior and exterior use, which dries to an eggshell finish or a soft semi-gloss finish	• Brush, roller or spray • Paint brushes and equipment to be cleaned with water	• Decorative finish coat for all surfaces • Used where there is poor ventilation (in toilets, kitchens, etc.) as it has low odour and is non-toxic • Requires no undercoat and dries quickly so that a second coat can be applied when required
Fire-retardant paint	• The main purpose of a **fire retardant** paint is to stop flame and fire spreading over a given surface • Some paints do this by releasing a flame-dampening gas once they become hot; other types have an **intumescent** property that contains a substance that reacts when exposed to a significant increase in temperature, leading the intumescent coating to swell	• Depending on type, can be applied by brush roller and spray and water-borne coatings can be cleaned with water	• Can be used on a wide range of surfaces, both internally as well as externally, to increase the surface protection from the effects of fire
Stains (interior and exterior)	• Acrylic wood stains can be used on interior and exterior timbers. When applied they soak deep into the timber surface to emphasise the grain of the wood • Available in a variety of colours, from natural wood shades to vibrant colours intended to change the appearance of the timber • Can be sealed with clear varnish or polish once fully dried	• Brush, roller or spray • Paint brushes and equipment to be cleaned with water	• Exterior and interior timber • Decorative internal timbers • Decking, fences, sheds, etc.

Coating type	Description	Application and cleaning	Uses
Preservatives (exterior)	• Preservatives are usually solvent-borne coatings which are used for wood treatment for smooth or rough-sawn exterior timber	• Brush or spray • Paint brushes and equipment to be cleaned with white spirit	• Preservatives kill and prevent wood-boring insects and prevent wet and dry rot • Re-coatable in 24 hours
Varnishes (interior and exterior)	• A varnish is a transparent liquid that is applied to a surface to produce a hard, protective transparent coating • Water-borne and solvent-borne varnishes are available • Varnish may be clear, but it is also available already stained to imitate different wood colours • When applying varnish, the object is to produce an even level film, free from runs, sags and with no dust or bittiness	• Brush • Brushes and equipment to be cleaned with white spirit if solvent-borne coating used and with water if water-borne coating used	• Can be used both internally and externally • It is important to apply the varnish firmly and confidently • Previously varnished surfaces should be lightly rubbed down to de-nib, then dusted off and the surface wiped over with a tack rag
Emulsion varnish (interior)	• A milky-white material that provides a clear washable surface when dry. It is grease and food-stain resistant • Can be thinned using water • Washable and resistant to mild chemicals and is non-toxic	• Brush, roller or spray • Paint brushes and equipment to be cleaned with water	• Used as a protective coating on wallpaper
Polyurethane varnish (interior and exterior)	• Clear surface coating available in gloss, matt or eggshell finish • If you need to thin it, use the manufacturer's recommended solvent • Water, chemical and heat resistant • Good adhesion	• Brush or roller • Paint brushes and equipment to be cleaned using manufacturer's recommended solvents (instructions vary)	• Used for protecting new and stained timber • Used to protect paintwork, furniture and special decorative finishes such as marbling and graining • Although hard-wearing, it is not really suitable for exposed exterior surfaces
Oil–resin varnish (external)	• A liquid coating which, when dry, becomes a clear and protective film • Hard-wearing and suitable for external use as it is water and weather resistant • Dries to a high-gloss finish	• Brush or roller • Paint brushes and equipment to be cleaned with solvent (white spirit)	• Used for protecting new and stained timber • Used to protect paintwork, furniture and special decorative finishes such as marbling and graining
Quick-drying varnish (interior)	• A fast-drying, high-quality varnish that is easy to apply and has a very low odour • During application it has a milky-white appearance, but when dry it forms a clear finish • Available in high-gloss or satin finish	• Brush, roller or spray • Paint brushes and equipment to be cleaned with water	• Gives good protection and decoration for interior timber and re-coating a previously coated surface that is in good condition
High build wood stain (exterior)	• A highly durable **micro-porous**, translucent, semi-gloss finish which comes in a variety of colours (wood tones) • Will form a very flexible film once dry that can withstand changes in timber without cracking (contracting or retracting)	• Brush or roller • Paint brushes and equipment to be cleaned with solvent (white spirit)	• The flexible micro-porous properties of high build wood stain make it particularly suitable for the protection and decoration of exterior timber surfaces such as window frames and doors

Coating type	Description	Application and cleaning	Uses
Universal preservative (exterior)	• Although universal preservative is similar in consistency to stains and varnish, it is a clear liquid and is solvent-borne • Contains fungicide • It takes a relatively long time to dry (16–24 hours) under normal conditions	• Brush or roller • Paint brushes and equipment to be cleaned with solvent (white spirit)	• Usually applied to new softwood that has not been treated with a preservative • Suitable as a coating for weathered timber surfaces • Stir well before use and apply one generous coat, paying particular attention to the end grain and joints
Protective wood stain (interior and exterior)	• A specially formulated protective wood stain • Not to be used on painted or varnished timber	• Brush and lint-free rag • Paint brushes and equipment to be cleaned with solvent (white spirit)	• Can be used on both exterior and interior surfaces, on softwood and hardwood as a decorative treatment • Apply two coats of the wood stain by brush and lint-free rag • Allow to dry overnight between coats

Types of tools used for applying coatings

There is a wide selection of brushes and rollers on the market which are used to apply paints, stains and clear coatings. The following will give you an understanding of which tool to use for a particular job. It is important to buy good-quality tools and equipment to apply surface coatings as cheap items will affect the finish.

Brushes

There are five parts that make up a brush:
- **Handle**: Usually made of a hardwood such as beech and sealed to make handling and cleaning easier and to stop water soaking into and damaging the wood. It can also be made from plastic.
- **Ferrule/stock**: This is a metal band that holds the filling and the handle together.
- **Epoxy/setting**: An adhesive which cements the filling by its roots into the stock.
- **Spacer**: A small wood, plastic or cardboard strip that creates a reservoir to carry paint.
- **Filling**: Usually natural bristle or synthetic man-made hairs, such as nylon.

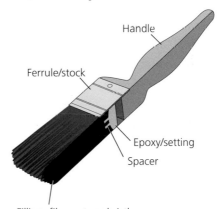

Handle

Ferrule/stock

Epoxy/setting

Spacer

Filling: filaments or bristles

▲ Figure 4.26 The parts of a brush

KEY TERM

Cutting in: the process of producing a sharp, neat paint line between two structural components in a room, such as a wall and ceiling or architrave and wall.

Types of brush

Natural bristle brushes are usually made from pig, hog or boar hair and are particularly suitable for applying oil-based paints. Synthetic (man-made) bristles are springier and are better suited to applying water-borne paints.

Paint brushes, whether natural or synthetic, come in different sizes for different jobs. The table on the right gives examples of brush sizes used for specific tasks, but as you gain experience you may develop a preference for a particular size or type of brush.

Brush size	Uses
12 mm	Used for difficult-to-reach areas, e.g. between two architraves on wall areas or for cutting in around window panels, skirtings, angles and edges of mouldings
25 mm	Can be used as above and for general cutting-in work, e.g. the edges of walls and ceilings when using rollers. This brush is also known as a sash tool
50–75 mm	For applying paint coatings to medium-sized areas and doors, etc.
100 mm	For applying paint (usually water-borne) to flat areas such as small ceilings and walls
125 mm and upwards	For applying water-borne paints to large surface areas and applying adhesive to wallpaper

The table below shows some of the different types of brush you may come across and use within the trade.

Type	Description and use
Flat paint or varnish brushes 	• Available in pure bristle or synthetic hair versions, the cost of these brushes varies according to the quality and the quantity of the filling • They can be used for applying most types of paint and varnish coatings to a variety of surfaces, including doors, window frames, ceilings and wall areas
Washing-down brushes 	• Relatively cheap two-knot or flat brushes, available in one size only and used for washing down with sugar soap or detergent
Flat wall brush 	• Available in a wide range of varying qualities and either man-made or pure bristle. The quality is dependent on the weight and length of the filling • Used to apply emulsion to large flat areas, e.g. ceilings and walls, and also to apply adhesive to wallpaper

Type	Description and use
Two-knot brush 	• Available mainly in pure bristle. The knots are usually bound in copper wire, as it does not rust • Used to apply water-thinned paints to rough surfaces such as cement, rendering and brickwork. They are also used to apply cement-based paints, as the bristles are not attacked by alkali in the cement, and for washing down surfaces when using a cleaning agent such as sugar soap
Cement paint brush (block brush) 	• These have man-made filling or coarse white fibre that has been set in a polished wooden handle. Cheaper block brushes are available in plastic • They are inexpensive brushes for applying masonry finishes and cement paints to a rough surface such as cement, rendering or brickwork
Fitch brushes 	• Available with pure bristle or synthetic filling, which is usually white, set in a round or flat ferrule • These brushes are used for fine, detailed work in areas that are difficult to reach with a paint brush
Radiator brush 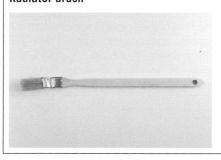	• These have a bristle filling attached to a long wooden handle or a wire handle that can be bent to fit into awkward areas • Used to apply paint to areas that are difficult to reach with a paint brush, particularly behind pipes, radiators and columns

IMPROVE YOUR ENGLISH

List the different-sized brushes that are used at your college or training centre and describe the areas and components that each brush could be used for.

Rollers

A paint roller is an application tool used for painting large flat surfaces rapidly and efficiently. It typically consists of three parts:

- a roller frame or cage
- a yoke that is attached to the roller frame or cage that forms the handle
- a roller cover or sleeve.

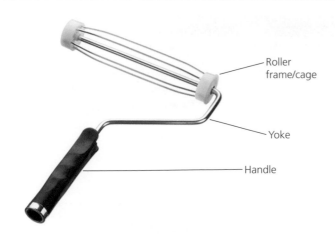

Roller frame/cage

Yoke

Handle

▲ Figure 4.27 Parts of a paint roller

The roller cover absorbs the paint and transfers it to the painted surface. The roller frame is reusable.

Applying paint to a large flat surface may be quicker using a paint roller. Specially shaped rollers are also available for painting corners, but sometimes it can be easier to use a paint brush.

The standard type of roller used by decorators is a cylinder roller which consists of a straight cylinder with a fabric cover or sleeve.

Choosing a roller will depend on the type of coating being used and the type of substrate to be painted. The different types include very smooth for applying finishing paints to flat doors, and lambswool for applying paint to textured surfaces such as pebbledash – it is important that you select the appropriate roller for the job.

When working on ceilings or high walls an extension pole attached to a roller may reduce the need for scaffolding.

The following table shows some of the different types of roller you may come across and use within the trade.

Type	Description and uses
Mohair roller	● Rollers made from natural mohair are very expensive, but synthetic mohair rollers are available that are more affordable ● Short-haired rollers are used for applying gloss paint to a smooth surface, medium-haired rollers are used for applying emulsion and long-haired rollers are used for pebbledash
Short-pile lambswool rollers	● Lambswool roller sleeves are made from the wool of a sheep and are used to apply water or oil paint to a smooth surface such as plaster, plasterboard or metal
Long-pile lambswool rollers	● These have a much deeper **pile** which is suitable for applying water-borne paints to brickwork and pebbledashed surfaces
Woven long-pile rollers	● All woven rollers are made of synthetic filaments. Long-pile rollers are used mainly for applying emulsion and masonry paint to pebbledashed surfaces ● Woven rollers come in 330 mm widths, so become very heavy when loaded with paint. All woven rollers are very similar to lambswool rollers, but they are much cheaper and can be thrown away after use
Woven medium-pile rollers	● These are used for applying emulsion, primer, rust-protection paint and varnish to small surfaces, or to semi-rough surfaces
Polyester long-pile rollers	● These are synthetic fabric rollers that have a highly absorbent 18 mm pile ● Used for applying water-borne coatings such as emulsion and masonry paint to rough areas

Type	Description and uses
Polyester medium-pile rollers	• These do similar jobs to the long-pile ones but have a 12 mm pile
Woven short-pile rollers	• The 6 mm deep pile is used for applying emulsion, primer, rust protection paint and varnish on small surfaces
Microfibre rollers	• Various sizes and depths of pile. Ideal for applying water-borne paints. These rollers typically pick up and release more paint than other types of roller pile
Small rollers/mini rollers	• Small 100 mm rollers can also be bought in long-, medium- and short-pile versions, and can be made from a natural or synthetic material • Used for applying paint to small areas such as flush doors (those with a smooth back and front), door panels, door furniture and small wall areas. They can also be used for applying paint behind a radiator

To make the task of applying coatings to large areas easier and quicker, a roller scuttle or roller bucket can be used. This holds more paint than a roller tray and is easier to move around when in use.

▲ Figure 4.28 Roller scuttle/bucket

An **extension pole** will enable you to reach higher areas such as the tops of walls and ceilings, although it can also be used to apply coatings to walls at normal heights and widths. Due to the distance that you can stand away from the surface, the coating can be applied in one fluid roll of the roller, giving better application and resulting in less overspray.

▲ Figure 4.29 Roller pole in use

IMPROVE YOUR MATHS

Choose a suitable roller and brush for painting doors, medium-sized areas and large areas. Then cost a range of different makes, sizes and fillings of rollers and brushes, starting with cheap items and finishing with more expensive items.

Record the cheapest and the most expensive then work out the range, the median and the mean of the price range.

Types of coating application defects

As a professional decorator, you will always aim to produce a perfect finish, but you may still make mistakes that need correcting. In Chapter 3 you looked at common defects that may occur when preparing previously painted surfaces. These and other defects are considered here in more detail, with a focus on their causes and how to avoid creating them in the first place. The table below describes the different application defects that you may come across, or create yourself, and how to rectify them.

KEY TERMS

Strain: to pour a coating through a porous or perforated device or material in order to separate out any lumps or bits.

Obliterate: to fully cover up/obscure.

Application defects	Description	How to rectify defect
Bittiness 	• Caused by the coating not being strained, or by the surface not being dusted down properly after rubbing down • It may also be caused by other trades people carrying out tasks around you or just walking by if there is dust on the floor	• If the coating is the problem, stop and **strain** the coating as the problem will not just go away • Dust off the surface if necessary and make sure you don't apply coatings while someone is sweeping up around you • Bits and nibs on the surface will have to be thoroughly rubbed down once completely dry before the next coat is applied
Misses 	• Areas which have been missed when applying paint, generally through carelessness	• When dry, the area will have to be re-coated and then checked to make sure the area is uniformly covered. If not, the full area will need to be re coated
Grinning	• Occurs when a coat of paint is too thin, i.e. it has not **obliterated** the surface to which it has been applied • This may be due to the paint being applied unevenly or too thinly, or the dramatic colour difference between the surface colour and the new colour	• You may have to apply several coats before an even finish is created

Application defects	Description	How to rectify defect
Runs and sags	• Caused by the over-application of a coating, which at first sags and then turns into runs before drying • Coatings need to be applied evenly and laid off so that the coating flows into itself	• To rectify these defects, wait until the coating is thoroughly dry, not just touch dry • The surface will then need to be wet and dried before re-coating
Excessive brush marks	• This defect is where the laying off brush lines can be seen after the final laying off process has been completed • When laying off, use light brush strokes so the coating flows into itself	• To rectify this defect, wait until the coating is thoroughly dry, not just touch dry. The surface will then need to be rubbed down before re-coating • Try not to break through the surface of the coating film when rubbing down, or you may have to undercoat the area again
Ropiness	• Also known as **tramlines.** This occurs when the coating does not flow evenly. This is usually caused by faulty workmanship • May occur as a result of applying the coating unevenly or over-brushing the coating until it starts to set (not keeping the edge wet)	• As above
Fat edges	• An application fault in which a thick ridge of coating occurs on a corner or **arris** (sharp external edge, e.g. of a door) • It can be avoided by laying off at the corners with an almost dry brush • This can be a problem when painting doors, as paint tends to build up on the edges	• If you produce a fat edge, you will need to wait until it has fully dried and then start the preparation process again, ready for another coat • Remember to apply the coating as stated
Wet edge build-up	• See 'fat edges'	• See 'fat edges'
Paint on adjacent surfaces	• Can be due to being over-vigorous with brush strokes or because the brush was overloaded • Make sure that surrounding areas are protected before starting work	• Remove splatters with the appropriate solvent – water for emulsion, white spirit for gloss paint
Roller edge marks	• Caused by applying uneven coating to the roller during application and the edges of the roller marks not being sufficiently laid off	• Reapply coating during application process and roll out coating evenly
Roller skid marks	• Usually caused by overloaded rollers skidding over the surface during application when applying too much pressure to the roller or roller pole	• Do not put too much pressure on the roller during application and allow the roller to apply the coating evenly

Application defects	Description	How to rectify defect
Irregular cutting in	• Poor cutting in skills can result in irregular lines that will need repainting to produce a neat finish. This is particularly noticeable when using two different colours	• Wait until dry and then reapply coating by cutting in correctly and neatly
Flashing	• Flashing is the presence of glossy patches on a painted surface • It can be caused by applying paint over an unprepared surface which leads to uneven sinkage due to different absorption rates • Another cause can be the wet edges setting before the joins are overlapped – this mainly occurs on ceilings and walls • Using cheap coatings can also cause flashing	• Good preparation of surfaces and correct application techniques will reduce the likelihood of many defects occurring • Applying coatings over well sealed and undercoated surfaces and using quality products can reduce the likelihood of flashing • Reducing the room temperature can also help solve this issue

Post-application defects

The following table describes the defects that can occur after coatings have been applied.

Post-application defect	Causes	Things to avoid when applying the coating
Retarded drying	• This is when paint takes a long time to dry. This is caused by **humidity**, which retards evaporation of solvents from the paint • Paint drying can also be impeded if it is colder than 10°C	• Direct sunlight will dramatically increase the paint temperature (and thus the speed of drying), with dark colours absorbing heat much faster • Wind and air movement speeds drying because as air passes over the wet paint it helps the solvents in the paint to evaporate
Cratering	• Craters in the surface of a dry paint surface can be caused by rain, condensation or heavy dew falling onto the wet paint surface before it is dry	• Avoid painting exposed surfaces if rain is forecast. Avoid painting in damp, humid atmospheres • Never apply paint to an external surface if it does not have time to become touch dry before dew starts to rise
Bleeding	• Bleeding occurs when the applied paint is stained or discoloured by the previous coating • Substances likely to cause bleeding are bituminous coatings and residues, tobacco tar deposits and resinous materials in timber. Burst water pipes or overflows can also cause bleeding which will stain surfaces • Bleeding may not become evident until some time after painting has been completed; this usually happens when timber has been pre-primed before use and the knots have not been sealed	• Where a potential cause of bleeding, e.g. a wallcovering or bituminous coating, can be removed before painting, it is advisable to do so • Alkali-resisting primer is effective in preventing bleeding from residues of tobacco tar (nicotine staining), which may remain after the surface has been washed thoroughly with sugar soap in the absence of a more specialist primer • Shellac knotting is commonly used to treat knots in timber to prevent them from bleeding • Any shellac-based coating can be used to seal bleeding and prevent further discoloration

Post-application defect	Causes	Things to avoid when applying the coating
Blooming	• This is a whitish appearance on the surface of varnish and can sometimes be accompanied by loss of gloss	• Avoid water being absorbed into the paint coating – always ensure that the surface is dry before applying coatings • Avoid applying coatings in humid conditions and try to avoid applying the coating in cold, damp conditions • Pollutants within the atmosphere can also cause this defect. If you suspect that pollutants have come into contact with the surface, you should wipe it over with white spirit and allow it to dry
Loss of gloss	• A gradual loss of gloss is to be expected as a finish ages • Early loss of gloss may be caused by applying coatings in unsuitable conditions, e.g. low temperature or high humidity, or to surfaces which are contaminated with grease, oil, wax, polish or hairspray, etc. • If the surface has not been sealed before applying coatings, or poor-quality coatings have been used, this may result in sinkage of the finish and the loss of gloss	• If the applied coatings dry satisfactorily and are not otherwise affected, application of a further coat will usually restore the gloss finish
Fading	• Loss of colour on a painted surface due to exposure to sunlight, ageing or exposure to weather • Colours that fade with the action of sunlight are known as **fugitive colours**. They tend to fade more in flat finishes than when protected by a gloss medium	
Discoloration	• Discoloration of coatings is often a result of the effect of atmospheric pollutants on ingredients in the paint • Exclusion from natural daylight may cause yellowing of coatings containing drying oils, while exposure to bright sunlight may result in fading of some pigments • Some types of moulds or fungi can also cause discoloration of coatings	• When discoloration has occurred, there is usually no alternative but to re-coat the surface • If a recurrence of the defect is to be avoided, it is necessary to establish its cause and, if possible, to use materials resistant to the conditions
Yellowing	• Where paint, usually white, gradually yellows over time • It is caused when linseed oil or phenolic resin-based (i.e. oil-based) paints receive little or no light, for example if they are behind pictures or furniture, or inside cupboards	• These areas will require redecoration with a non-yellowing paint coating
Cracking/crazing	• Occurs in paintwork and is usually due to the application of a hard-drying coating (oil paint) over a softer coat (paint that has not fully dried) so that the top coating is unable to keep pace with the expansion and contraction of the previous coat as it dries	• If cracking has formed on the surface there is no remedy except to burn off the defective paint

Post-application defect	Causes	Things to avoid when applying the coating
Flaking/peeling	• Where the paint film lifts from the surface and breaks away in the form of brittle flakes • It is caused by applying paint to moist or loose, powdery surfaces	• Remove the defective surface down to a smooth finish and then ensure that the surface is dry and stable before coating

KEY TERMS

Humidity: moisture in the air.

Fugitive colours: colours that fade when exposed to light. Some colours that are reasonably stable when used at full strength develop fugitive tendencies when mixed with white to create a lighter shade.

Storage and cleaning of tools and materials

As with all tools and equipment, there is little point in buying good-quality paint brushes and rollers if they are not cleaned and stored properly at the end of each job or task.

When cleaning brushes, rollers and other equipment used such as scuttles and work pots, you need to know which type of cleaning solution to use so you can clean the items correctly and not cause further damage to them.

Tools and equipment that have been used for oil-based/solvent-borne coatings will need cleaning with a different solvent from tools used for water-based/water-borne coatings.

Cleaning brushes

When using brushes, remember to remove all remaining coating from the filling and stock of the brush (see Figure 4.31).

To remove coatings correctly:
1 Identify the type of coating that has been used and whether it was oil-based/solvent-borne or water-borne.
2 Pour any excess paint from the kettle or roller tray back into the paint container and wipe the brush on the container to remove as much of the coating as possible. Some decorators use a piece of board to wipe their brush on.
3 Wash brushes in the appropriate medium: for oil/solvent-borne coatings use white spirit or the manufacturer's recommended cleaner, and use water for water-borne coatings.

When cleaning oil-based/solvent-borne paint brushes, use the following steps:

1 Pour the cleaning agent (white spirit) into the paint kettle and use a vigorous pumping action to remove paint from the stock. Repeat until there is no evidence of any colour coming from the brush.
2 Spin the brush between your hands to remove as much of the solvent as possible (paint brush spinners can help with this job).
3 Wash in warm soapy water until a clean lather is obtained and then rinse thoroughly in clean water.

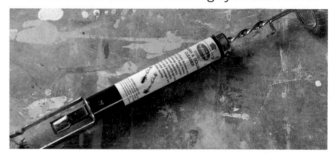

▲ Figure 4.30 Paint brush spinner

When cleaning out water-borne paint brushes, use the following steps:
1 Follow the same process as for oil-based paint.
2 Rinse in cold water until the water runs clear.
3 Then wash in warm soapy water.
4 Rinse in clean water.
5 Mop up excess water with a rag before storing.

▲ Figure 4.31 Clean off water-borne paint with warm soapy water

Cleaning rollers

Cleaning rollers takes much longer than cleaning paint brushes, and the longer the pile the more time consuming the process. Using cheaper roller sleeves that can be thrown away at the end of a job can be more cost effective, but the environmental costs must also be considered.

It is important to consider the time it will take to clean roller sleeves, as this can have a significant effect on the time it takes to complete a job. Painting a door with gloss paint may seem quicker with a mohair roller but you will also need to use a brush for cutting in, and at the end of the process the mohair roller will need scrupulous cleaning in large amounts of white spirit.

The process for cleaning rollers is similar to that for cleaning paint brushes, but remember to use white spirit for oil-based/solvent-borne coatings and water for water-borne coatings.

When rollers are clear of all coating residue, remove as much moisture as possible by spinning the roller.

Some colleges and industrial sites use a waste management system such as Safetykleen that provides a constant flow of solvent to clean brushes, rollers and equipment quickly and efficiently. It is essential to wear gloves and goggles when using this type of equipment.

If you cannot clean up your brushes and rollers at the end of the day they can be **steeped** in water or solvent until you can clean them properly. Do not leave them too long, as the liquids will evaporate and the brushes and rollers will dry out, making them unusable.

For short breaks you can wrap rollers or brushes in a plastic carrier bag or plastic film to keep the surface moist.

KEY TERM

Steeped: soaked in liquid.

INDUSTRY TIP

Always check that tools are clean before use. Kettles, buckets and roller trays should be wiped out before use to remove dust particles.

HEALTH AND SAFETY
Remember to wear rubber gloves and goggles to protect your hands and eyes from paint and solvents.

Cleaning other equipment

To clean other equipment used during the application process, such as paint kettles, roller trays and roller buckets/scuttles, wipe any excess paint back into the paint container and wash the equipment with the appropriate solvent.

Storage of equipment

Once items have been cleaned and dried they can be stored in a dry area, as damp may cause metal to rust and timber to rot.

Clean paint brushes should be stored in a cool, dry place. Excessive heat may cause the bristles and setting material to shrink, resulting in loose ferrules and bristles. Damp conditions can cause mildew to develop on the bristles and damage them. Never store brushes with the filling downwards, as this will bend the bristles and ruin the brush.

A **brush keep** or brush keeper can be used to store solvent-borne brushes in a wet state. It works by solvent being contained in a bottle with an evaporating wick. The fumes from the evaporating solvent replace the air in the brush keep, preventing the brushes from drying out.

▲ Figure 4.32 Brush keep

Rollers should be stored in moderate temperatures where they cannot be contaminated by chemicals, oil or grease. If possible, hang them up so that air flows around the roller sleeve.

If you cannot hang up rollers, stand them upright, as storing them on flat surfaces may crush the pile of the roller.

Store paint and materials in accordance with COSHH data sheets

The materials used by painters and decorators need to be treated carefully to ensure everybody's safety, and this section looks at ways to keep risks to a minimum. In order to work safely with substances that can be hazardous to health, it is important to know what the risks are.

Storage of coatings

COSHH (Control of Substances Hazardous to Health) data sheets are available for the substances you will be using. The data sheets provide the following information:

- what the product is
- what it is used for
- identification of hazards
- information on the composition of ingredients
- first aid measures
- firefighting measures
- accidental release measures
- instructions for handling
- guidance for storage.

A coatings data sheet will give you all the information for that particular coating. The information should be on the label of the product, but if it is missing or the product has been decanted into another container the data sheets will be available online or from the manufacturer.

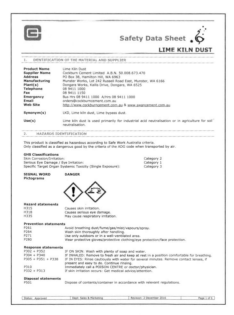

▲ Figure 4.33 Example of a COSHH data sheet

▲ Figure 4.34 Example of a paint safety data sheet

A coatings data sheet will give you all the information required for the use of that particular coating. This information should be on the label of the product, but if it is missing or the product has been decanted into another container the data sheet will be available online or from the manufacturer.

> **HEALTH AND SAFETY**
>
> Use the internet to research two-pack epoxy solvent floor paint. Find the relevant COSHH data sheet and write a short report on how to use the product, health and safety considerations, including first aid, and transport and storage instructions.

Disposal of waste

Correct disposal of waste after applying coatings is very important. Care must be taken when disposing of hazardous waste and it should never be poured down the sink. Try not to buy more paint than you need for a job and consider applying another coat if you have some left over.

Empty emulsion tins must be washed out and can then be disposed of in household waste.

It is now possible to buy paint solidifiers, which are small beads that absorb the paint, turning it into a solid mass that can be disposed of in the household waste.

Check with your local authority for how to dispose of oil-based paints and varnishes.

Rags and cloths that have been used to apply chemical solvents should be fully opened out, allowed to dry and then disposed of carefully, as they can be a fire risk if left bound up in a wet state as **spontaneous combustion** can occur.

INDUSTRY TIP

It is not advisable to decant hazardous substances into new containers. Always read the instructions for storage and if you have any leftover product that cannot be stored, follow instructions for how to dispose of it.

KEY TERM

Spontaneous combustion: an internal reaction causes an increase in temperature which results in a fire starting without any outside influence.

Storing protective sheeting

After using cotton dust sheets and other protective sheeting, give them a light shake once finished with (although not near wet paint!) and fold them up ready for reuse. If sheets are wet or damp, they should be allowed to dry before being folded and stored to prevent mildew forming on them. All protective sheeting should be stored on shelves off the floor. It is important to keep sheeting clean and dry because it will be used to protect floors and furniture another time. Send sheets away if they need cleaning.

Protective sheeting such as tarpaulin, PVC-coated nylon, rubber-coated cotton and heavy cotton canvas should be stored in the same way as cotton dust sheets. Rubber tarpaulin can be wiped clean using a sponge and warm water and allowed to dry before being folded and stored.

If protective sheeting is not used for some time it will need to be checked for damp. If there are signs of damp, unfold and air-dry it before re-folding and storing.

▲ Figure 4.35 Plastic tarpaulin sheet

Storage of materials, tools and equipment

New or partially used coatings, materials, tools and equipment will need to be stored correctly for safety and to ensure that they do not **deteriorate**. They should be stored in such a way as to protect your own health and safety and that of others. Many chemicals used by decorators have a low **flash point**, and if they are kept in conditions that are too hot, they are likely to explode or catch fire.

INDUSTRY TIP

Never smoke around highly flammable materials.

KEY TERMS

Deteriorate: to become worse.

Flash point: the temperature at which a material gives off a vapour that will ignite if exposed to flame. Chemicals with a low flash point are labelled as highly flammable.

The storage area should be dry, well ventilated and frost-free all year round. Make sure that there is no risk of materials coming into contact with naked flames such as gas heaters, boilers, etc., as many of these materials are flammable.

The storage area should be fitted with sturdy racking, with large and heavy materials stored on the bottom shelves. Never store powder filler or textured finishing materials on concrete floors, as they can remain cold and damp even in warm weather and will be unfit for use. Small containers of filler that have been opened can be stored in airtight plastic containers.

HEALTH AND SAFETY

Be careful how you lift items from storage areas. Always use safe handling techniques.

INDUSTRY TIP

Always check the manufacturer's instructions for storage.

Oil-based/solvent-borne coatings (undercoat, gloss and varnish) and water-borne coatings (emulsions and masonry paints) should be stored on shelves and clearly

173

marked with the labels turned to the front and used in date order.

Oil-based/solvent-borne materials should be **inverted** at regular intervals to prevent the pigments settling and the ingredients separating. Check that the lids are on firmly first.

Some water-borne paint products such as emulsions and acrylics have a limited shelf life and should be used before their use-by date.

Fattening is where the paint in partially filled containers becomes very thick, making it hard to apply. The paint should either be thinned before use (but remember the danger of the pigment not providing sufficient coverage) or disposed of.

When receiving a delivery of new paint materials, make sure that the old stock is brought to the front to be used first and the new stock stored behind it. This is known as stock rotation, and it helps to ensure that old stock is not left to go out of date.

Some paints are susceptible to a defect called to **livering**, where the paint thickens to a jelly-like condition (like raw liver) by oxidation during storage.

Two-pack paints are specialist paints used for cars and aircraft which dry to an extremely hard finish. You may also come across a two-pack epoxy solvent floor paint. These paints are extremely toxic and must be handled with care. Ensure that they are stored upright with the lid firmly closed. The paints must be kept away from heat and naked flames.

> **HEALTH AND SAFETY**
>
> Never stack materials so high that there is a danger of them falling. Never over-reach yourself if you are trying to get a product from a high shelf – you don't want it falling on you or covering you from head to toe in paint!
>
> Appropriate fire extinguishers should be on hand in case of a fire.

> **IMPROVE YOUR ENGLISH**
>
> Research fire extinguishers and find out the correct type of fire extinguishers that should be used for the different materials used for painting surfaces and substrates.

> **HEALTH AND SAFETY**
>
> Make sure that lids and caps are closed tightly to limit the escape of VOCs into the air.

Apply coatings by brush and roller

When applying coatings to surfaces, substrates and areas, it is important to follow the correct sequence of application to avoid making mistakes as this will result in a poor-quality finish.

You need to be aware of the task sequence when applying coatings to make sure that you keep the edge of the paint wet and that you do not mark your completed work or spill paint onto it. If you are working in an area inhabited by people, you may need to adjust your schedule to work around them.

Preparing coatings for use

Before you apply coatings, you first need to prepare them for use. After the surfaces have been prepared and the paint system has been chosen, you will need to get the coating from a larger container and **decant** it into a paint kettle, roller tray or roller scuttle/bucket.

> **KEY TERMS**
>
> **Inverted:** turned upside down.
> **Decant:** to transfer a liquid by pouring from one container into another.

If the container of paint/coating has not been used before, you may need to remove any dust from the lid of the paint container. You will need to open the lid using a paint tin opener; never use the edge of a paint scraper or filling knife, as the blades are easily damaged.

You will then need to stir the paint with a paint stirrer or palette knife until all the sediment is dispersed and the required consistency is achieved.

▲ Figure 4.36 Paint stirrers

Pour the required amount of paint into the paint kettle or roller tray. Using your brush, remove any paint that may have gathered in the rim of the paint container and then wipe clean using a rag. Replace the lid of the paint container so that the remaining paint does not become contaminated.

If the container of paint has been used previously, remove any dust from its lid and open it as above.

The air trapped inside the container when the lid was last replaced may have caused a skin to form on the surface of the paint. If there is a layer of skin present, it can be removed by cutting the skin away from the edge of the inside of the container. Lift out the skin intact if possible and dispose of in a waste bin.

▲ Figure 4.37 Old paint tins need to be opened with care

Search the paint for lumps and debris by straining. Place a paint strainer on a paint kettle and pour the required amount of paint through the strainer to remove any bits of skin or contamination present from the last time the container was opened, then remove the strainer and clean or dispose of it as appropriate. Clean the rim of the container as before and then replace the lid.

Traditional strainers will need cleaning after each use, or they will clog up when the paint dries. You can purchase single-use disposable strainers, but these are more expensive, so making them yourself is a cheaper option and is quite easy to do (see Industry Tip).

INDUSTRY TIP

You can use old tights or stockings to strain the paint and then dispose of them afterwards; this is cheaper than purchasing disposable strainers.

The **viscosity** of the coating must be checked to make sure it is the correct thickness to apply to the surface. Coatings that are too thick are hard to apply and leave fatty edges and excessive brush marks, while coatings that are too thin will not give sufficient coverage to obliterate the previous coating and colours may bleed through.

Check the manufacturer's instructions before thinning out coatings and remember that when using water-borne coatings you thin with water and when using solvent-borne you use white spirit.

KEY TERM

Viscosity: the ability of a liquid or coating to flow; the more viscous it is, the slower it flows.

Using a paint kettle

The aim of a decorator should be to transfer coatings to surfaces without covering themselves or spilling them everywhere. A paint kettle makes this easier because of its weight and the fact that the opening on the paint kettle is large. Although you should always wear protective overalls, you should endeavour to work as cleanly as possible, as wet paint has a habit of transferring itself to places where it should not be. Follow these steps to minimise paint transference:

1 Only pour as much coating as is needed into the paint kettle.
2 Use only one side of the kettle so you have a clean side and a wet side.
3 Load the brush with paint/coating and use the side of the kettle to remove any excess.
4 When breaking from work, lay the brush across the top of the kettle with the bristles lying on the paint/wet side of the kettle so the handle rests on the clean side.
5 If you are stopping for a longer period of time, stand the brush in the paint kettle with the handle against the clean side.
6 When you resume work, wipe off the excess coating (on the paint/wet side of the kettle) before continuing with the task at hand.

Painting large areas

When painting large areas, such as ceilings and walls, the following must be considered:

- **What is needed to reach the work areas?** If you are working on a ceiling or a high wall you may need some form of scaffolding to reach the work area, or you may be able to manage with a pair of steps. The important thing is to plan how you are going to reach the whole area to be covered, because once you have started you will need to keep the edge going so that the paint will flow into itself and not leave fatty edges.
- **Is the surface flat or textured?** This will determine what tools and equipment you will need to carry out the painting.
- **What is the drying time of the paint or coating?** The manufacturer's instructions on the back of the container will tell you how quickly the coating will dry. If the drying time is quick and you have a large area to paint, you may need a second person to help. The instructions will also tell you how long you will need to wait before giving the surface a second coat.
- **Is the work area to be painted manageable by one decorator or will it require more than one person?** Remember that you will need to keep the edge wet to ensure you produce a solid paint film by eliminating brush strokes. You may need to work with a partner to achieve this.
- **Is the surface porous or non-porous?** This may affect the drying time, the consistency and the amount of paint required.
- **What should you use to apply the paint?** For small areas such as doors, window frames and pipes, you may need only a brush, but for larger areas you may need a roller for the large area and a brush for cutting around the edges. If two are working on the same area one may cut in and the other may apply the paint using the roller.
- **Where should you start?** When painting large areas, plan where to start and where you will finish to ensure that the edge does not dry off before the next application of paint. Look at the area that is going to be painted and remember that you will need to keep the edge wet, so whether it is a ceiling, wall or door, always start at the narrowest part.

Apply paint coatings to ceilings and walls

The following information explains how to decorate a typical domestic room. You should be familiar with applying coatings to straightforward flat areas and

by following these sequences you should be able to achieve an overall good standard of finish:

Always start decorating a room by painting the ceilings first, working from one end of the room to the other. When cutting in it is advisable to extend the paint 15–25 cm on to the top of the wall to give you a nice line to cut in to. Once the coating has dried the ceiling can be re-coated.

▲ Figure 4.38 Painting a ceiling

The wall areas are painted next. Apply the coating to one wall at a time. If you are using brushes, you may need a 25 mm brush for cutting in and a 150 mm brush for the rest. If you are using a roller and brush, cut in a straight line where you have overlapped the paint from the ceiling and then paint the edges before filling in with the roller. Do not re-coat until the first coating has fully dried.

When you are painting an area with a large brush or roller it can be difficult to get into the corners and around obstructions. Before you start you will need to use a small brush to make a neat line around door frames, windows, mouldings and internal angles. Professional decorators rarely use masking tape to cut in around windows and other obstacles, as it is time consuming and paint sometimes seeps under the tape anyway. You will develop the skill to paint straight lines neatly, although it may take a while to have the confidence to paint freehand. Practice makes perfect!

Once you have cut in the edges you can then fill in the area with a larger brush or roller.

▲ Figure 4.39 Cutting in

Once the ceilings and walls have been coated, you then move onto the doors, windows and **linear** work (skirting boards, door frames, picture frames and dado rails). Most will require two coats.

KEY TERMS

Linear: relating to straight, often narrow, lines.
Contaminate: to pollute or infect.

INDUSTRY TIP

Never use paint directly from the manufacturer's container, as any **contamination** can ruin the whole batch of paint or coating.

Water-borne coatings

When applying water-borne coatings such as emulsion paint to ceilings or walls by brush, work in stages as illustrated below. However, do not lay off in the conventional way, as you would when painting a door, for example (see later in this chapter). Cross-hatching is the best process to use when applying emulsion, to minimise the effect of brush marks (tramlines) created by the brush.

When matt emulsion dries, the light is **refracted** and does not run down the brush marks, thus making the paint appear more opaque and matt.

KEY TERM

Refract: to deflect light from a straight path.

Vinyl silk emulsion highlights defects of both the surface and the brush marks, but it is easier to keep clean and can be wiped over with a cloth.

If you are using oil-based paint on ceilings you will need to follow the same process as for applying a coat of paint to walls.

In order to keep the wet edge to a minimum in terms of both time and area you will need to work on small areas at a time by mentally dividing the wall into small squares.

This illustration shows the sequence for one decorator applying paint to walls:

5	3	1
6	4	2

One decorator paints sections 1–6. If two decorators are working on a large wall, one would cut in while the other decorator filled in with a roller.

This illustration shows the sequence for two decorators working together applying paint to a ceiling:

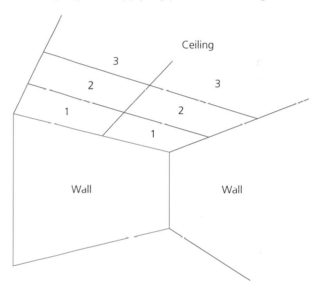

One decorator paints sections 1–3 on the right-hand side and at the same time the second decorator paints their sections 1–3 on the left-hand side. In this way a wet edge is maintained.

As before, keep the edge wet. The size of the brush you use will depend on the area that is being painted, but remember that you are aiming to keep the edge wet and apply an even coat.

When using brushes and rollers on large surfaces such as ceilings and walls remember the following points:
- First cut in the edges at the ceiling or wall line.
- Then cut around obstacles such as electrical fittings and any fixtures.
- Always use a suitably sized paint brush for the cutting in.

Apply oil-based coatings by brush to non-complex areas

The sequence below describes how to apply oil-based paints to non-complex areas.

KEY TERM

Laying off: using the very tips of a brush to lightly stroke the surface of the paint or scumble to minimise brush marks.

Apply oil-based paints by brush

STEP 1 Apply the first application of coating to the surface using the cross-hatch method. Work in areas of approximately 300 mm square along the surface and then continue down or across the surface. You will find that you make mistakes when applying the paint, for example putting it on too thickly or not brushing it out evenly, but don't worry – the more you practise, the fewer mistakes you will make.

STEP 2 The next step requires you to lay off the applied paint in the short direction, overlapping each brush stroke by a third of the width of the brush. The paint will flow into itself and make an invisible join.

STEP 3 When applying the paint, you will need to apply light pressure to the brush to make the bristles work the paint to an even application.

STEP 4 The final stage requires you to lay off in the final direction, lengthways. You do not need to load any more paint onto the brush for the **laying off** process.

ACTIVITY

Take a flat piece of timber approximately 700 mm × 1000 mm that has been primed and practise this method of applying oil-based paint before attempting it on a real door. Use different pressures to lay off the paint. Once dry, check to see which pressure produces the best finish and the least noticeable brush strokes.

INDUSTRY TIP

Close windows and doors when applying paint to large areas. The lack of ventilation will give a longer wet-edge time and help to slow down the drying time because the evaporation rate of the solvents will be reduced. Wear a respirator when doing this, as some paints may give off harmful fumes.

Apply coatings to doors

Here are some factors to consider when applying oil-based/solvent-borne and acrylic/water-borne coatings to doors:

- Is it an external door? External doors are often made from hardwood to help protect them from the weather and should be coated with an oil-based/solvent-borne coating/paint. Remember that the coating/paint will dry quicker outside.
- Is it an internal door? Internal doors are more likely to be made from softwood and will not be affected by bad weather, so they can be painted using either oil-based/solvent-borne or acrylic/water-borne coating.
- Is the surface flush or panelled?
- Is the surface PVC or metal?
- Are there any glazed (glass) areas?

The traditional sequence of painting a flush door starts with painting the edge of the door, followed by the top left-hand corner and ends at the bottom right, as shown in Figure 4.40.

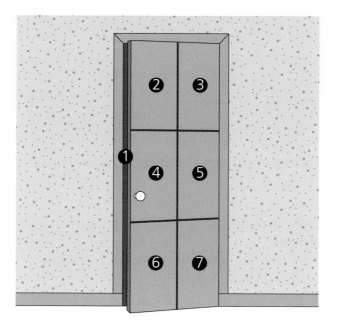

▲ Figure 4.40 Painting a flush door

If a door opens towards you into the room, you will need to paint the edge with the lock and fittings on. When painting a door that opens away from you, you will need to paint the hinge edge.

Paint the edge before the face of the door, as once the face has been painted there may be a build-up of paint on the edge, which will be less noticeable when the paint has dried.

Remember: Always rub down between coats to remove nibs and so on, and then use a dusting brush to remove any fine dust remaining on the surface.

It is quite common for flush panelled doors to be painted with a roller and then laid off with a brush to remove the roller effect. This can make for a speedier process and reduces the likelihood of uneven application and dry edges.

When painting or varnishing a panelled door follow the sequence in Figure 4.41, starting at the top left panel to help prevent fatty edges forming. Remember to lay off following the direction of the grain.

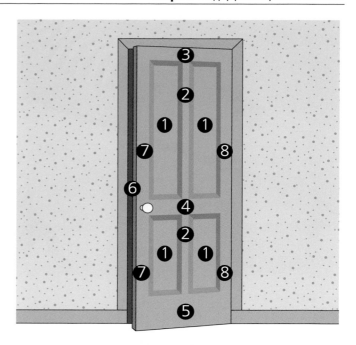

▲ Figure 4.41 Painting a panelled door

INDUSTRY TIP

Panelled doors consist of horizontal members ('rails') and vertical members ('stiles' on the outside of the door or 'muntins' in the centre). Always start with the mouldings, then panels (1), then top rails (3), lock rail (4) followed by muntins (2) then door edge (6) followed by stiles (7/8) and finishing with the bottom rail (5).

All rooms have some form of linear work as stated above, consisting of some or all of the following surfaces:

- door frames
- skirting boards
- mouldings
- dado rails
- picture rails.

When painting these surfaces, use a small cutting-in brush and make sure that the surfaces are not overloaded with paint, as this could result in runs and sags, spoiling the finished look. These complex areas require careful painting, as they are often decorative and will really stand out in the completed room.

▲ Figure 4.42 Painting a skirting board

Apply coatings to windows

Most new builds have PVC windows fitted, and PVC is also used to replace old decayed timber windows or rusty metal windows in some properties. However, you will still come across many different types of timber or metal windows that need painting.

As there are so many different designs, all with different types of openings, window painting is one of the most time-consuming jobs when decorating properties both internally and externally.

▲ Figure 4.43 Bay window

There are many different designs of windows and it can take some time to master the sequences for each one, but the overall principles are the same.

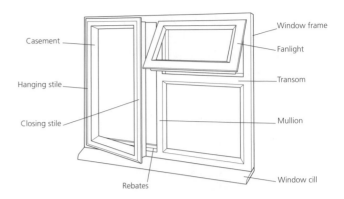

▲ Figure 4.44 Component parts of a casement window

Painting standard opening windows

The following steps show the painting sequence used for standard opening windows:

1 Open the windows and paint all the frame **rebates**.
2 Paint fanlights, cutting in around the puttied areas.
3 Paint the hanging stiles.
4 Paint the closing stiles.
5 Paint the mullion.
6 Paint the transom.
7 Paint the window frame.
8 Finish by painting the window cill.
9 Allow to dry fully before closing the windows.

KEY TERM

Rebate: a rectangular area removed from the corner of a timber section. Used to locate a pane of glass while a frame rebate locates a window sash.

Painting a sash window

The following steps are required when painting a sash window. First paint the internal window as follows:

1 Slide the outer sash down a little and raise the inner sash, leaving a gap at the top and bottom.
2 Paint the glazing bars and the surface of the inner sash, including the top surface of the meeting rail and the underside of the bottom rail.
3 Paint the parts of the outer sash that are visible.
4 Slide the outer sash until it is almost closed and lower the inner sash (grip it from the outside so you don't touch the wet paint).
5 Complete the painting of the outer sash, omitting the top rail.

6 To paint the pulley stiles, pull the sash cord away from the surface so that you can paint behind it. Avoid painting the sash cord, as this will hinder the smooth running of the sash window. Paint all remaining liners and shutters.

7 Finish by painting the cill.

Follow the steps below when painting the external window:

1 Reverse the sashes by sliding the outer sash up and the inner sash down.

2 Paint all visible surfaces except the top surface of the top rail. The pulley stiles can be painted at the same time if not already painted.

3 Return the sash windows to an almost closed position and complete the painting of the inner sash and the pulley stiles.

4 Paint the cill.

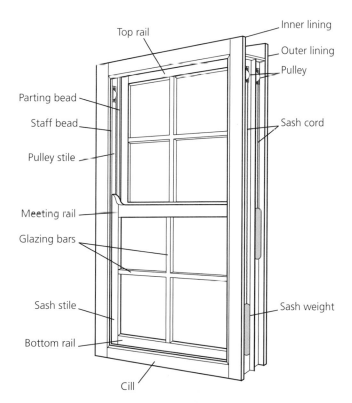

▲ Figure 4.45 Component parts of a sash window

INDUSTRY TIP

Never start painting windows if you think that they will not be dry enough to close before you leave the job, as the paint films will bond together and they will not open.

Apply coatings to a staircase

Begin by covering the floor with dust sheets and secure them with masking tape. Once the surfaces have been prepared, follow the sequence below.

1 Apply the coating to the balusters (also known as spindles), starting from the top.

2 Then apply the coating to the outer string.

3 Then apply coating to the newel post.

4 Then apply coating to the handrail.

5 Finally, apply coating to the wall string skirting.

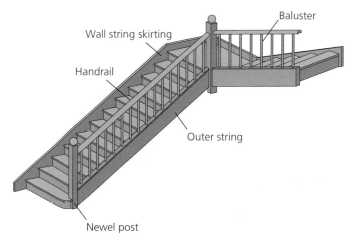

▲ Figure 4.46 Painting a staircase

IMPROVE YOUR MATHS

You have been asked to buy paint to cover a 6 m × 5.5 m room that has four walls 2.5 m high. Calculate the total area and how much paint you will need in order to give the walls two coats, if a litre of paint covers 16 m². How much paint will you need to buy, and how much will it all cost if a litre of paint costs £10.55?

(Answer provided on page 301.)

Apply coatings to ferrous and non-ferrous surfaces

Before applying coatings to ferrous and non-ferrous surfaces, check if the surface needs priming. Use the appropriate primer followed by an undercoat and a finishing coat. The method for applying paint and coatings is the same for all surfaces.

1 Lightly rub down with fine abrasive.

2 Dust off.

3 Apply the coating and lay off to form an even coat of paint.

Apply coatings to timber surfaces

Some varnishing or staining of timber surfaces is classed as a specialist job, but you may be required to varnish a timber surface such as a door or a skirting board when carrying out tasks during your career.

On surfaces that are to be covered with opaque paint the undercoat and primer will not show through the finishing coats, but this is not the case with translucent or transparent coverings. Stains and varnishes are used to protect timber surfaces without obscuring the beauty of the grain, so these surfaces are prepared in a different way from wood that is painted.

▲ Figure 4.47 A natural wood stain is used to emphasise the timber

Wood stains

Wood stains can be used on exterior and interior timbers and, when applied, they soak deep into the timber surface to emphasise the grain of the wood. They come in a variety of colours, from natural wood shades to vibrant colours intended to change the appearance of the timber. They can be sealed with clear varnish or polish after application.

Before applying the wood stain, make sure the surface is dry and then lightly rub down using fine **silicon carbide paper** (also known as wet and dry paper). Remember to rub down the timber with the grain. Once all the rubbing down is done, remove the dust and then apply the stain with a brush or a lint-free cloth.

Varnish

Varnish is a transparent liquid that is applied to a timber surface to produce a hard, protective transparent coating. Both clear and coloured varnishes are available and the choice will depend on the finish to be achieved.

When applying varnish to a surface, the aim is to produce an even level film free from runs, sags and **pinholing** and with no dust or bittiness, as these defects will spoil the overall finish.

It is important to apply the varnish firmly and confidently to the timber surface. If the coating is applied too thinly and is bare in places, it will be impossible to obtain an evenly distributed film and this will result in runs and a poor appearance.

KEY TERM

Pinholing: a pore-like penetration that is present in paints and coatings due to moisture, air, solvents or other fluids being trapped.

Previously varnished surfaces should be lightly rubbed down to de-nib, then dusted off and the surface wiped over with a tack rag. Knots should not be sealed with knotting solution as they will show through the varnish. Some surfaces will require more preparation and will need sanding, then wet abrading using silicon paper to achieve a better finish.

▲ Figure 4.48 Applying varnish to a timber surface

Test your knowledge

1 What statement best describes a domestic area?

    a An area where people work

    b An area where people play sports

    c An area where people live

    d An area where people are painting

2 VOCs stands for:

    a Very orange colours

    b Voice over commands

    c Very organic compounds

    d Volatile organic compounds.

3 Where is a low-tack masking tape most suitable for?

    a Exterior garden furniture

    b Carpets and flooring

    c Sign writing and stencils

    d All of the above

4 Identify the disadvantage of using cotton twill dust sheets.

    a When first used they give you a professional look

    b Paint spills may soak through them

    c They are available in different sizes

    d They move very little when in use

5 What does the 'S' stand for in the acronym DIPS?

    a Sanding

    b Staining

    c Safety

    d Sanitation

6 What kind of finish does the binder element of a water-borne coating give?

    a Gloss

    b Flat

    c Eggshell

    d All of the above

7 Identify the **first** stage of paint drying.

    a Coalescence

    b Oxidation

    c Evaporation

    d Abrasion

8 Where is matt emulsion most suitable for?

    a Walls and ceilings

    b Windows

    c Doors

    d Skirting boards

9 What part of a paint brush holds the handle and filling together?

    a Setting

    b Ferrule

    c Filament

    d Spacer

10 At what stage of the drying process when applying solvent-borne coatings can a second coat be applied to the surface?

    a Soft dry

    b Touch dry

    c Hard dry

    d Thorough dry

Practice assignment

You have been asked to carry out a refurbishment job for a domestic dwelling. The clients have had a double extension added to their existing home. The extension has newly plastered walls and ceilings as well as new softwood skirting boards, doors and architraves. They have also asked your opinion on what to coat the new oak timber staircase with. The extension has been added to existing rooms at the rear of the house, so you must also prepare existing surfaces in these rooms.

Prepare a report that identifies the correct tools, equipment and materials needed for the whole task. You also need to select the correct coatings for all the surfaces described above.

CHAPTER 5

APPLY PAPERS TO WALLS AND CEILINGS

INTRODUCTION

Applying paper is a fundamental skill required for any decorator to work professionally in both the domestic and commercial sectors. It is a skill that requires precision and methodical working, incorporating calculations, measuring and an eye for detail. Understanding paper types, patterns and their respective applications will provide the modern decorator with an invaluable skill and increase their employability status.

In this chapter you will gain the skills and knowledge required to plan and apply various standard papers to ceiling and walls. With practice, all these skills can become second nature.

By the end of this chapter, you will have an understanding of:
- paper production and application
- adhesives for wallpaper application
- applying standard papers to walls and ceilings.

The table below shows how the main headings in this chapter cover the learning outcomes for each qualification specification.

Chapter section	Level 1 Diploma in Painting and Decorating (6707-13) Unit 119	Level 2 Diploma in Painting and Decorating (6707-22/23) Unit 217	Level 2 Technical Certificate in Painting and Decorating (7907-20) Unit 205	Level 2 NVQ Diploma in Decorative Finishing and Industrial Painting Occupations (6572-02) Unit 338
1. Understand paper production	N/A	1.1–1.5	1.1–1.3	This is an optional unit and is not a mandatory requirement for achieving the NVQ. The skills and knowledge acquired, however, will enable a decorator to work on a much broader range of contracts
2. Know adhesives for wallpaper application	2.1–2.3, 3.1–3.6, 4.1–4.4	2.1–2.4, 3.1–3.4	2.1–2.2	
3. Apply standard papers to walls and ceilings	5.1–5.8, 6.1-6.8, 7.1–7.3	4.1–4.15, 5.1–5.10, 6.1	3.1–3.6	

1 UNDERSTAND PAPER PRODUCTION

Paper production methods

This section looks at some of the different methods of wallpaper production you may come across.

Making the paper

Paper is made from wood pulp and manufactured in a paper **mill**. **Synthetic** (i.e. man-made, not natural) fibres are sometimes added to give the paper additional texture and strength. A roll of paper from the mill is 1.65 m wide and typically over 6000 m long. This material weighs about one ton.

▲ Figure 5.1 Paper production

Before transferring to the printers, each roll is cut into 530 mm widths that are 10 m long. This is a typical size for a standard roll of paper. Rolls of greater length can be provided if required.

This method is used to produce a single layer paper (called **pulp**) that would be suitable for printing as a **simplex** paper.

For more complex papers such as **ingrains**, **duplex**, embossed, blown vinyl and vinyl-coated papers, there are additional processes that produce the desired texture or structure prior to the printing process.

Ingrains, duplex and embossed papers are manufactured using two layers of paper.

KEY TERMS

Mill: a place where paper is produced in its plain form before sending to the manufacturer for processing, in this case to be made into wallpaper.

Pulp: colour printed directly on an untreated paper (no ground colour applied). Pulp is a cheap form of simplex wallpaper produced in its simplest form from wood pulp. It is often a thin, inexpensive printed wallpaper. Pulps are single layer, without the luxury of protective layers such as vinyl, and are therefore easily marked.

Simplex: a wallpaper made from a single layer of paper. It can be produced in smooth or embossed patterns.

Ingrain: commonly referred to as woodchip. Chips of wood are sandwiched between two layers of paper.

Duplex: papers made of two layers for greater strength and to bond them together before being sent for printing. The layers are usually embossed to form a pattern in relief.

ACTIVITY

1 What is paper made from?

2 Is there a process for recycling wallpaper? Find out if it can be recycled in your area. Hint: in the UK, your local council is responsible for waste disposal.

Embossed papers

▲ Figure 5.2 An example of embossed paper

Embossed papers usually consist of two layers of paper. The paper passes through soft rollers (see Figure 5.1): one of the rollers has the pattern raised up on its surface, and the other has the same pattern recessed. As the paper passes through, the pattern is physically pressed into it: this is called a raised or relief pattern.

Anaglypta and duplex embossed papers are produced in this way and can be printed with colour or designed to be left white so that the decorator can paint over the paper later, when on site. Lower-cost embossed papers can be produced using a single layer of paper, but they are thinner and the pattern is lower relief (not as deep), so they tend to flatten when hung.

Supaglypta or **high-relief panels** are produced by the wet embossing method and have cotton lint and other fillers added between the layers of paper before pressing.

This method allows for a much heavier quality product that should ensure that the raised pattern remains in place after hanging. The hanging of this type is discussed in further detail at Level 3.

Heat expansion

Blown vinyl wallpaper is produced using the heat expansion method. An expanding agent is added to liquid vinyl (PVC), which expands in size after it is heated at high temperatures, producing a three-dimensional effect. When heat is applied to the PVC during the production process, the PVC on the flat wallpaper expands into patterned rollers to create the blown vinyl in a range of patterns and textures. Vinyl is printed on a paper substrate.

Further manufacturing processes are discussed in the next section on wallpaper printing methods.

▲ Figure 5.3 An example of blown vinyl paper

Printing processes and methods

The use of wallpapers probably dates back at least as far as the medieval period. At that time, wallpaper was a cheaper alternative to decorations such as tapestries, silks or wood panelling. Early wallpapers were block printed by hand onto paper squares, and then matched on the wall. Once machine printing was introduced, it became possible to manufacture wallpaper in long lengths, similar to those used today.

▲ Figure 5.4 Pattern printing six-stage build-up

The following information and illustrations provide some understanding of the printing processes used to produce the wonderful range of pattern, colour and texture found in modern wallcoverings.

Various printing methods can be used to add colour, texture and pattern to wallpaper. To print a pattern, a series of pattern plates is developed. Each plate contains one colour or one part of the pattern, and once each part has overprinted the one before, the complete pattern is visible on the paper. Each layer of the pattern must be dry before the next one is printed on top. Figure 5.4 shows how a pattern was printed in six stages (known as print runs). In this printing method, the first coating of colour to be printed is called the **ground**.

> **KEY TERM**
>
> **Ground:** the first coating applied when producing printed wallpaper – a colour that is applied to provide a background.

Block printing

Figure 5.5 shows the carved printing blocks, while Figure 5.6 shows the printer applying the block to the paper. The printer colours the block by lowering it down onto the colour tray. Once it is sufficiently inked it is lifted and manoeuvred over to the paper by an arch lever system (crane) before being pressed down to create the print. Pins on the side of the blocks guide the printer in placing the block exactly into position.

▲ Figure 5.5 Wooden printing blocks, each with a unique reference number

▲ Figure 5.6 Printing in progress with the block moved into place with the help of a crane

Accuracy and strength of colour are effectively governed by the printer's experience of how much pressure needs to be applied to the back of the block.

After each individual colour has been laid, the paper is hung for 4–5 hours to let the thick water-based inks dry, before the next colour is applied. Once all the colours have been printed, the wallpaper is PVA lacquered for protection, before being manually trimmed and individually hand wound.

Surface printing by machine

The surface print machine is the oldest of the mechanised processes and is the predecessor of the relief printing flexo machine. Invented in 1839, this method continued as the only mechanised means by which to print wallpaper for the next 100 years. It was responsible for making wallpaper available to the masses and its immediate success marginalised block printing.

The print cylinders are made of a very hard ceramic-type rubber and the area that is not for printing on is cut out from it, leaving the printing surface 'proud' on the cylinder. Figure 5.7 shows a surface printing machine with its multiple rollers and many colours being printed at the same time.

▲ Figure 5.7 Surface printing machine

The inks are water-based and transferred via a rotating woollen or felt blanket. The blanket is soaked with ink at one end, while the other end is touching the back of the print cylinder. As the conveyor belt-type blanket rotates it picks up the ink from the tray and delivers it onto the back of the print cylinder; the ink is then impressed directly onto the paper as it is rotated. The amount of ink transferred onto the cylinder is determined by the absorbency of the blanket. The more porous the blanket, the more ink; the harder the blanket, the less ink.

Flat-bed screen printing (silk screen printing)

Traditionally, screen printing was called 'silk screen printing' because the stencil screens were made of silk. This term has carried over into modern times even though nylon is now more commonly used. Screen printing is a relatively simple process that produces wallpaper with a wonderfully rich depth of colour.

The screen is a rectangular frame with a fine polyester nylon woven mesh stretched across it. To create the design, it is first necessary to completely coat the screen with a photosensitive polymer. The stencil of the design is then placed flat onto the mesh before the screen is 'photo-exposed' under special lighting.

For printing, the screen is laid face down onto a long flat table that has the wallpaper stretched tight along the length. A typical table is 2 m wide and 31 m in length. The thick water-based inks are applied to the back of the screen and are drawn across the mesh using a rubber squeegee (like a car windscreen wiper blade), forcing the ink through the areas of open mesh in the shape of the design. Once the colour is laid, the screen is lifted and moved along to the next position, where the process is repeated. However, the process is labour intensive and demands constant vigilance. As a result, wallpapers produced in this way tend to be comparatively expensive and are usually aimed at the top end of the market.

▲ Figure 5.8 Screen printing table

▲ Figure 5.9 Screen printing frame

Rotary screen printing

Rotary screen printing is a relatively recent development of flatbed screen printing. It can print a continuous web of moving paper, upwards of 3000 m long, as opposed to the limiting length (normally 30 m) of flat screen-printed wallpaper.

The screen has a width of up to 68 cm and is between 64 and 100 cm in circumference, thus allowing relatively large pattern repeats. A typical rotary screen print machine has an in-line configuration with upwards of seven or eight print stations available, each printing one colour. The cylindrical printing screen itself is a very fine 'honeycomb' mesh. Rotary screen printing is similar in principle to the hand screen method, but is cheaper due to its continuous process of printing.

▲ Figure 5.10 Rotary screen-printing machine

Flexographic printing

▲ Figure 5.11 Flexographic rubber roller

Introduced into Britain in the early 1960s and similar in principle to (and a development of) surface printing, flexographic printing (commonly referred to as 'flexo')

is a relief-type print process that uses a relatively soft rubber print cylinder with a 'raised' printing surface. The area not to be printed is cut out of the roller, leaving the raised area for accepting the ink (similar to the way a hand-held rubber stamp is cut). The ink is transferred from the ink tray to the print roller via another roller, the purpose of which is to both even out and determine how much ink is transferred onto the cylinder.

▲ Figure 5.12 Flex printing machine

Rotogravure printing

Gravure printing, also known as **intaglio**, uses a hard engraved cylinder to transfer the image to paper, but unlike surface and flexo, the image is recessed (directed inward) instead of being raised. The ink collects in the recessed pockets and is absorbed by the paper as it passes over the cylinder. The deeper the colour desired, the deeper the recessed pocket and the more ink transferred. Most of the borders that duplicate the look of photography or realistic art are printed by the gravure method.

▲ Figure 5.13 Rotogravure print machine

KEY TERM

Intaglio: a design incised or engraved into a material.

Heat embossing printed vinyls

There are generally two types of vinyl: solid PVC vinyl and blown PVC vinyl.

Blown PVC is sub-categorised into two types: 'mechanically' blown vinyl and 'chemically embossed' blown vinyl.

The purpose of heat embossing a vinyl is to convert a smooth plastic wallcovering into a wallcovering that has a third dimension, i.e. a tactile 'feel'. This textured 'relief' (difference in thickness or height) may be something as smooth as a **light sand emboss** (which ensures the vinyl does not look shiny under light), or it could be a heavily textured **rough emboss** that has a visual and tactile effect that works as an effect in its own right, as well as a support to the print.

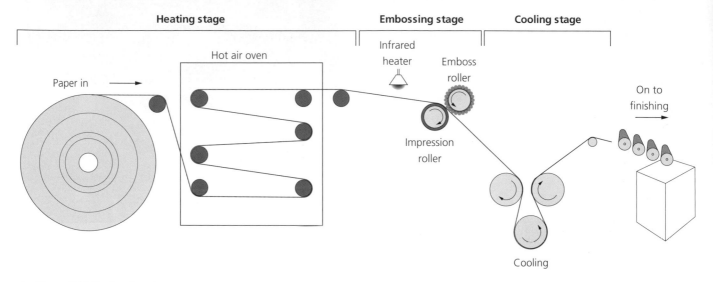

▲ Figure 5.14 Heat embossing process

Digital printing

The latest types of wallpaper printing machines are capable of printing up to 2.5 m wide and can be linked to the latest computer-based design programs. Bespoke designs can be produced on the computer and sent direct to the printer. Wide widths can be printed to fit clients' walls by matching each length as supplied by the printer. Figure 5.15 shows a digital print design station.

▲ Figure 5.15 Digital wide printer

Types of paper and their suitability for application

In this section we will consider in more detail the types of wallpapers that you need to know about at Level 2. Simplex, duplex and pulps are **generic** terms that relate to the initial production of wallpapers before the application of further treatments.

KEY TERM

Generic: members of a group or type rather than the name of a specific brand.

Foundation papers and finishing papers are used to produce various effects. Foundation papers can be used as part of the preparation process, while finishing papers create a visually pleasing finished effect.

Foundation or preparatory papers

Papers such as lining papers and ingrains are described as **preparatory** papers, and their primary use is to create a **defect-free** substrate on which to hang finishing papers or apply paints. Lining papers are used as a **foundation** for either painting or hanging finishing papers.

> ## KEY TERMS
>
> **Preparatory:** in the context of paperhanging, this refers to the preparation of a surface for further treatment.
>
> **Defect-free:** without flaws, holes or cracks and bits left on the surface.
>
> **Foundation:** in the context of paperhanging, this means a suitable base for further treatment, such as hanging finishing paper over the top or applying paint.

Cross lining is the process of applying lining paper horizontally as a foundation paper to a vertically hung finishing paper such as patterned paper. Hanging in this manner helps to avoid joints lining up in the layers of paper. Cross lining is also carried out where the finishing paper is delicate and to provide an even, non-porous base, particularly over areas of excess filling or high porosity.

Lining papers should also be used on gloss-painted walls, after suitable preparation, to enable the finishing paper to stick to the surface. Glossy surfaces offer poor **adhesion**, and without lining it is likely that the finishing paper will lift and peel away. The gloss surface should be abraded (preferably wet) to remove as much of the shine as possible. This abrasion will also provide a good key for the lining paper to adhere to. Hang the paper horizontally as a foundation for finishing papers, or vertically if you will be applying paints as the finish.

Silicon carbide abrasive paper is commonly referred to as '**wet and dry**' as it can be used as a wet or dry process. It is more effective when used wet and will produce a very smooth finish to the abraded surface.

> ## KEY TERMS
>
> **Cross lining:** the process of applying lining paper horizontally as a foundation paper to a vertically hung finishing paper such as patterned paper. This is carried out in some cases where the finishing paper is delicate and also to provide an even, non-porous base, particularly over areas of excess filling or high porosity.
>
> **Adhesion:** the action or process of adhering (sticking) to a surface or object.

Undecorated blown vinyl and embossed papers are also used as foundation papers, as they are usually painted over after hanging. Wood ingrains are also used in this way. These types of paper can be used to disguise rough and uneven surfaces, as their textured finish tends to disperse the light and avoids accentuating the defects beneath.

▲ Figure 5.16 Embossed paper can be hung and then painted

Finishing papers

These are used to provide an effect through a coloured, visually textured or patterned finish. Finishing papers include those listed earlier in the chapter, such as the various types of vinyl, embossed papers, printed pulps and grounds. Borders are also used to give a finished effect to the edges of areas and provide a visual impact or accent.

IMPROVE YOUR ENGLISH

Using an online dictionary or one from your library, look up the words 'accentuating' and 'accent' that appear in the sections headed 'Foundation or preparatory papers' and 'Finishing papers' on the previous page and explain their meanings in the sentences in which they are used.

Patterned and textured papers can provide a variety of designs and colours that enable you to achieve interesting and visually appealing effects in a way that plain colours do not. Finishing papers can change the appearance and visual dimensions of rooms. Patterns with a diagonal or horizontal emphasis can make rooms appear wider, while patterns with a vertical emphasis can make rooms appear taller.

▲ Figure 5.17 Vertical patterns make a room appear taller

▲ Figure 5.18 Horizontal patterns make a room appear wider

Lining paper

Lining paper is a white pulp paper sold in various grades such as 800 g/m^2 and 1000 g/m^2. The higher the number of grams per m^2 determines how heavy and thick the paper is. Typically rolls are 560 mm wide and 10 m long, although it is possible to buy double, treble and quadruple length rolls. Longer rolls can be more economical if carrying out a lot of lining as the waste is minimised.

Lining paper is used:
- to provide a uniform surface in terms of even **porosity** for the subsequent hanging of finishing papers
- to line non-absorbent surfaces such as oil painted walls, again to provide a surface of even porosity; hanging finishing papers direct to non-absorbent surfaces could lead to defects such as springing of joints, poor drying out leading to blistering and poor overall adhesion
- to provide even porosity on surfaces where large areas of making good have changed the overall porosity of the surface.

Hang lining paper using cellulose paste. For thinner varieties starch ether or all-purpose paste may also be used.

Non-woven lining paper

Fibre lining paper is a non-woven material that is most suitable for covering up unsightly and poor plaster surfaces. It is extremely strong, does not require any soaking and can also be hung directly on pasted walls. Use on old and poor plaster, breeze and cement blockwork, wood panelling and cement render, etc. A good-quality heavy-duty paste should be used for all types of surfaces, and the wall should be pasted rather than the material. All surfaces must be well prepared before hanging fibre lining, and when hung it can be emulsion painted, used as a base for overhanging other wallcoverings or as a base for specialist paint effects. Fibre lining paper is a replacement for the old cotton-backed lining paper.

To hang fibre lining paper, use the **paste-the-wall technique** with manufacturers' medium-weight ready-mixed **PVA adhesive**.

KEY TERMS

Paste-the-wall technique: the wall is pasted rather than the back of paper.

PVA (polyvinyl acetate): a resin used in both adhesives and paints to provide a hard, strong film. When used as an adhesive, the film is clear and does not stain.

Adhesive: in decorating and particularly paperhanging, an adhesive is a material sometimes referred to as paste that can stick paper to ceiling and wall surfaces.

Wood ingrain paper

Wood ingrain paper is often referred to as woodchip. This is a pulp paper of two layers, where small pieces of wood chips are sandwiched between the two layers. It usually comes in 10 m length by 530 mm wide rolls although, as with lining paper, it is possible to obtain double, treble and quadruple rolls for better value. Wood ingrain can be supplied in different grades of texture – fine, medium or coarse. The different grades lead to an appearance that is more or less pronounced. Wood ingrain is usually coated with water-based paints or sometimes oil-based paints after hanging. Ingrain papers tend to mask irregularities in the underlying surface due to the pronounced texture of the woodchip appearance.

As with lining paper you should not over-soak the wood ingrain paper when hanging as thinner grades may tear and possibly **delaminate**.

▲ Figure 5.19 Wood ingrain paper

KEY TERM

Delamination: the separation of the top and bottom layers of paper when over-soaked.

Hanging

Hang wood ingrain paper using starch or starch ether paste.

Embossed wallpaper

Embossed wallpaper was discussed earlier in the chapter. Embossed papers are typically produced by two different types of process: dry embossing or wet embossing.

Wet embossing produces the heavier high-relief patterns, and both methods mostly use duplex paper to produce the raised pattern. It can be produced in colour or left in the white for overpainting.

Hanging

Hang embossed paper using starch paste or starch ether. Do not overfill the relief and take care when brushing out to avoid flattening the emboss or relief.

▲ Figure 5.20 Embossed paper

Blown vinyl wallpaper

Expanded blown vinyl textured paintable wallpaper is created by an application of liquid PVC to flat wallpaper. There is a blowing agent within the PVC and when heat is applied during the production process, the PVC on the flat wallpaper expands to create the blown vinyl, which is textured, white, paintable wallpaper.

Expanded blown vinyl textured paintable wallpaper is used to disguise walls with cracks, lumps, bumps and even old woodchip wallpaper. The textured blown vinyl wallpaper can then be painted to create a tough, hard-wearing, textured wallcovering.

▲ Figure 5.21 Blown vinyl wallpaper

Hanging

Hang blown vinyl wallpaper using starch ether or all-purpose paste containing a fungicide. If blown vinyl is to be overlapped in corners, then overlap PVA adhesive must be used, as ordinary paste will not stick vinyl to vinyl. Bear in mind blown vinyl

overlapped will produce a rather bulky appearance so it may be best to splice the overlap to enable the bulky underlayer to be removed. (This is discussed further later in this chapter.)

Washable wallpaper

This wallpaper has a paper substrate on which the decorative surface has been sprayed or coated with an acrylic coating. These wallpapers are classified as scrubbable and strippable and are suitable in almost any area. Washable papers provide better resistance to grease and moisture than plain paper and are good for bathrooms and kitchens.

Hanging

Hang using starch ether or all-purpose paste containing a fungicide.

▲ Figure 5.22 Washable wallpaper

Vinyl wallpaper

Vinyl wallpaper has a paper (pulp) substrate laminated to a solid decorative surface. This type of wallpaper is very durable since the decorative surface is a solid sheet of vinyl. It is classified as scrubbable and peelable.

Paper-backed vinyl/solid sheet vinyl

Paper-backed vinyl consists of a solid vinyl layer of material laminated or bonded to a paper backing sheet. These wallcoverings have a heat-embossed (raised) effect to **register** (fit) the pattern design and provide a multitude of textural effects. They have a high-quality appearance and are durable because the decorative surface is solid vinyl. Vinyl wallpapers resist moisture, stains and grease and are scrubbable and peelable, but will not withstand extreme physical abuse. Cleaning may be more difficult due to the raised pattern.

Fabric-backed vinyl

Fabric-backed vinyl wallpaper has a substrate laminated to a solid vinyl decorative surface. It is generally considered the most durable wallcovering because the vinyl is a solid sheet and not applied in a liquid form. This type of wallpaper is ideal for consumers looking for a great degree of washability, scrubbability and durability. It commonly used in commercial settings.

Hanging

Hang vinyl wallpapers using starch ether or all-purpose paste containing a fungicide.

▲ Figure 5.23 Vinyl wallpaper

Ready-pasted wallpaper

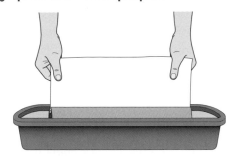

▲ Figure 5.24 Using a pasting trough for soaking

This vinyl wallpaper has been pre-pasted in the factory with a water-activated fungicidal paste. The cut length or roll is dipped in a trough of water to wet and activate the paste.

Hanging

After soaking the paper in a trough as per the manufacturer's instructions, it can be applied to the wall and smoothed down using a sponge.

Non-woven wallpaper

Non-woven refers to a combination of natural and synthetic fibres that provide a light paper backing that is easy to hang and easy to remove. The patterns are printed or embossed onto this and it is more difficult to tear, even when it is wet during the application process. The material does not expand or contract in use, making it easy to hang without the requirement for it to be soaked. The special blend of natural and synthetic fibres makes non-woven wallpaper breathable and washable.

▲ Figure 5.25 Non-woven wallpaper

Hanging

Paste is applied to the wall over an area slightly larger than the length of paper to be hung. Once the paper has been measured and cut to length it is hung in the normal way, using a spatula to smooth out from the centre to the edges and top to bottom.

If necessary, use a seam roller for the edges. Be careful with delicate surfaces and only use a roller with sponge rubber or foamed material. Do not apply too much pressure, especially when working on the edges, as the pattern can be distorted, which will cause problems when applying the next length or adjusting the pattern to the last length.

Once the wallpaper is applied make sure the face of the paper is sponged to remove any excess paste that may have been transferred to the face of the paper. The wallpaper is best trimmed using a trimming knife and straight edge. Continue hanging in the same way for adjoining lengths.

Borders

▲ Figure 5.26 Hanging a wallpaper border

Borders supply a further decorative feature to the look of a room and are commonly used between the ceiling and the top of the wall. They can also be used around frames or to make framed feature panels as well as being used at dado height. Fashion and current taste tend to dictate their use and positioning.

Borders are manufactured in rolls and generally supplied in 10 m lengths, while widths can vary greatly. As with wallpaper, an abundance of patterns and textures is available, and borders can add a decorative feature to plain painted areas. Figure 5.26 shows one such application.

Hanging a border

Borders are normally unrolled on the paste table and pasted using PVA adhesive. The border is **concertina**

folded (see Figure 5.27) and hung in the same way as cross lining.

KEY TERM

Concertina fold: wallpaper folded like the folds in a concertina musical instrument.

▲ Figure 5.27 Concertina folding enables small folds when working horizontally as in cross lining or hanging borders

ACTIVITY

Create your own record of paper types either by using images from the internet, or by collecting/cutting different samples from your college or training centre.

Pattern and matching papers

All wallpapers, except some textures and murals, have a **pattern repeat**. The repeat is the vertical distance between one point on the pattern to the identical point vertically. This pattern repeat is an integral part of the design. The repeat can range anywhere from 20 mm (or sometimes less), up to as much as the width of the wallpaper or more.

Understanding **pattern match** and pattern repeat is one of the more difficult areas of wallpapering for many. Understanding the pattern repeat is particularly relevant when calculating how much wallpaper is required, as buying too much can be expensive, whereas not buying enough could mean going back to the shop and finding that the supplier no longer has any of the same batch left in stock.

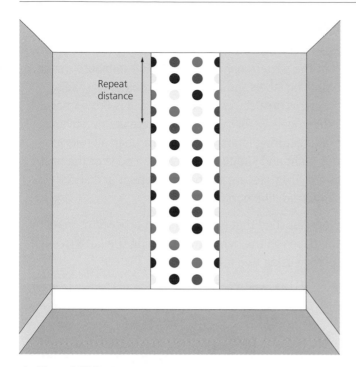

▲ Figure 5.28 Pattern repeat

KEY TERMS

Pattern repeat: the distance between a single point on the pattern and the next point where it is repeated on the pattern.

Pattern match: the pattern match helps you identify where the pattern at the edge of one piece of wallpaper fits together with another roll. This could be an offset match, such as a half-drop or random match, or a straight match. The type of pattern match is given by the symbol on the packet.

▲ Figure 5.29 Matching wallpaper pattern

If the wallpaper to be hung has a pattern, find out what type of pattern match it has. There are three major types of pattern matches:

- **Random match:** The pattern matches no matter how adjoining strips (strips that are next to each other) are positioned. Stripes are the best examples of this type of match. It is generally recommended to reverse every other length to minimise visual effects such as shading or colour variations from edge to edge. Note that any random match will produce less waste since there is no repeat distance to take into account.
- **Set/straight match:** This match has a pattern which matches on adjoining lengths. Every strip will be the same at the ceiling line.

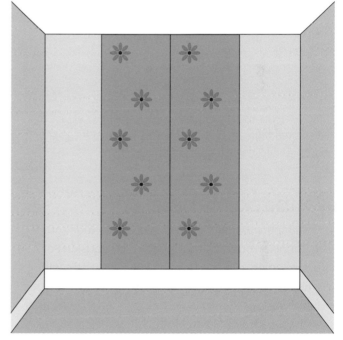

▲ Figure 5.30 Set pattern match

- **Drop pattern match or offset pattern match:** Every other length is the same at the ceiling line and the pattern runs diagonally. It takes three lengths to repeat the vertical design. Every other length is identical, so when cutting it is best to number the sheets on the back to avoid confusion. It is quite common to cut paper from two rolls when dealing with drop patterns to try to minimise the amount of waste produced. Drop patterns, particularly those with a large pattern repeat distance, always produce large amounts of waste when matching.

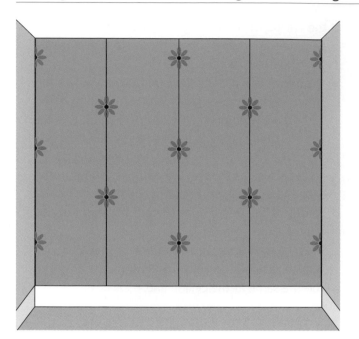

▲ Figure 5.31 Offset or drop pattern match

INDUSTRY TIP

Drop patterns usually require a greater number of rolls to be ordered than straight match. This is due to the greater waste produced when matching the pattern.

Wallpaper labels

All wallpapers will be labelled and will include some very important details that need to be considered before hanging. On the front of the label there will be the pattern number, **batch** number, shade number and also various symbols denoting whether the paper is washable, straight match, etc. More information on these symbols follows later in this chapter.

The back of the label will contain all the hanging instructions recommended by the manufacturer.

▲ Figure 5.32 Batch numbers and symbols on wallpaper rolls

Pre-hanging checks

Read the label carefully before opening the roll to check that the pattern numbers and **batch numbers** match. Matching labels should avoid any colour or shading problems. Visually check that the roll is not damaged. When opening rolls check that there are no obvious manufacturing faults. It is usual to open all rolls at this point and **shade** them over the edge of the paste board. Try to locate the pattern repeat and also the suggested direction of hanging.

Once satisfied that all is well then proceed to match the paper on the bench by matching the pattern edge to edge using two rolls.

▲ Figure 5.33 Visually check wallpapers and colours

KEY TERMS

Batch number: an identification number used to denote when a batch of wallpaper was produced. It will contain a code indicating the print details. Codes should be the same on all rolls to avoid colour differences.

Shading: visually checking the shade or colour of wallpaper rolls. Batch and shade numbers should be the same on every roll.

▲ Figure 5.34 Pattern matching on the bench

International performance symbols

The following symbols can be found in a wallpaper pattern book or on product labels on rolls of wallpaper. Use this guide to help identify the various characteristics of the rolls to ensure the best possible results.

Symbol	Characteristics
Spongeable	• Either printed with waterfast colours or thinly coated with PVA • Paste on the surface can be safely removed by gentle use of a damp sponge
Washable	• Like spongeable, but with more protection • Will withstand more wiping after light soiling
Super washable	• Like spongeable, but well protected with PVA • Suitable for wet areas such as bathrooms and kitchens
Scrubbable	• Normally only applies to vinyl wall coverings used in areas with heavy wear
Good lightfastness	• This means that the wallcovering should retain its original colour for a considerable amount of time, but cannot be guaranteed on any permanent basis
Overlap and double cut	• Where edges cannot (or should not) be butt-jointed in the usual way, they are lapped and the two thicknesses sliced through, usually with a straight edge and a knife. The subsequent two strips of waste are pulled away, leaving two edges stuck down as a perfect **butt joint** • Often applies to contract materials and thicker wall coverings
Peelable	• The top skin of the decoration should peel cleanly away, leaving the backing (if still secure) ready for further decoration

Symbol	Characteristics
Offset match	● This is sometimes called a half-drop or drop pattern – the match is obtained by halving the repeat
Straight match	● The pattern matches straight across the width
Free match	● No matching required – lengths can be cut straight from the roll without wastage
Moderate	● Beware! This product may well fade or discolour in areas of direct sunlight
Adhesive to wallcovering	● This needs to be pasted, as it is not a ready-pasted or paste-the-wall product
Ready-pasted	● To be hung after immersion in a trough of clean water according to the instructions
Paste the wall	● Applicable only to Novamura, some textiles and some wide-width wall coverings ● Always check the type of adhesive and the method of application
Strippable	● Should peel cleanly from the wall in a dry state

Symbol	Characteristics
Design repeat distance offset $$\frac{50}{25}\,\text{cm}$$	● The circumference of the printed rollers, half of which represents the wallcovering drop (where applicable) ● This is very important information for calculating the number of lengths available from a roll and will give the total quantity you will need
Direction of hanging ↑	● This can occasionally be found on the selvedge of wall coverings that have been untrimmed, or it may be on the reverse of some products. When shown on a label it means 'don't reverse the lengths'
Reverse alternate lengths ↑↓	● Mostly applicable to plain effects and necessary to minimise the risk of side-to-side shading
Co-ordinated fabric available	● This does not necessarily mean an exact matching fabric even when the design is the same, as colours may vary because of the different printing techniques
Wet removable	● This will not dry-peel

KEY TERM

Butt joint: edges of lengths of paper that touch without a gap or overlap.

IMPROVE YOUR ENGLISH

Find a wallpaper information label online or at your college or training centre. Copy the information about the name of the paper and the shade or batch number. Also copy the symbols relating to this specific wallpaper and write down what they mean.

ACTIVITY

1 What is a batch number?
2 What is shading?

IMPROVE YOUR ENGLISH

Find a wallpaper information label online or at your college or training centre. Copy the information about the name of the paper and the shade or batch number. Also copy the symbols relating to this specific wallpaper and write down what they mean.

2 KNOW ADHESIVES FOR WALLPAPER APPLICATION

Select and prepare adhesives

A variety of adhesives are available. Adhesives are considered in terms of their water content, as this has the most impact when choosing an adhesive for a particular surface or paper.

Types of adhesive

The types of paper studied at Level 1 and Level 2 are relatively lightweight, and the following types of adhesive are suitable for use with them:

- **Starch paste**: This is typically supplied in powder form and is based on organic starches extracted from maize, corn, wheat and so on. It is easy to prepare and should be used when freshly made. It can be used on any type of paper as long as it contains fungicide. If no fungicide protection is indicated, it is best not to use it for vinyl papers because of the potential for **mould** growth under the surface of the paper.

- **Cellulose paste**: This will generally produce the thinnest of pastes and is therefore only really suitable for lightweight papers such as lining papers and other lightweight foundation papers.

- **Cellulose paste moderated with starch ether**: This is suitable for all types of paper, including lightweight vinyls. It usually contains fungicide to reduce the likelihood of mould growth.

- **Ready-mixed adhesive (medium weight)**: This is typically a PVA-based adhesive. It can be supplied in light, medium and heavy grade depending on the type of wallpaper to be hung. It is designed to roller or brush directly from the tub for speedy application – no mixing or dilution is required. The super-smooth, easy-spread adhesive has strong grab and easy slide for perfect wallpaper hanging.

- **Border and overlap paste**: This is a PVA resin-based adhesive supplied in tubes and plastic tubs. It is commonly used for sticking overlaps of vinyl wallpaper, as well as fixing borders.

- **PVA**: As described above, ready-mixed adhesives and border/overlap adhesives use PVA resin to provide a strong bond with the surface. Papers hung with PVA are often quite difficult to remove because of this strong bond, but its use is essential when hanging heavyweight materials.

KEY TERM

Mould: mould is made up of airborne spores that can multiply and feed on organic matter in pastes (starch pastes contain organic products such as wheat). Mould typically shows as black spots on the surface of paper.

INDUSTRY TIP

Should mould occur on walls and ceiling surfaces, then this will need to be treated with a fungicidal wash before removal and careful disposal to avoid spreading to other surfaces.

The table below shows the advantages and disadvantages of common wallpaper adhesives.

Type of adhesive	Water/solid content	Advantages	Disadvantages
Starch paste	Medium solid, low water	<ul><li>Relatively slow setting</li><li>Adheres well to surfaces</li></ul>	<ul><li>Easily stains the face of the paper</li><li>Encourages and supports mould growth if not fungicide protected</li></ul>
Cellulose paste	Low solid, high water	<ul><li>Fairly transparent and therefore less likely to stain the face of papers</li></ul>	<ul><li>Has less adhesion than starch paste</li><li>When cellulose paste has been standing around for a long period of time, it will become thin and unusable</li></ul>
Cellulose paste modified with starch ether (all-purpose paste)	Medium solid, medium water	<ul><li>Fairly transparent</li><li>Mixes easily to a smooth paste</li><li>Better adhesive properties than cellulose</li></ul>	<ul><li>Can mark the face of papers more than cellulose</li></ul>
Ready-mixed adhesive (medium weight)	High solid, low water	<ul><li>No mixing</li><li>Easy to apply</li><li>Suitable for hanging all types of lining papers and wallpapers</li></ul>	<ul><li>More expensive than other types of adhesive</li></ul>
Paste-the-wall adhesive	High solid, low water	<ul><li>No mixing</li><li>Apply direct to wall</li><li>Suitable for hanging all types of non-woven wallpapers</li></ul>	<ul><li>More expensive than other types of adhesive</li></ul>
Overlap adhesive	High solid, low water	<ul><li>Ready mixed</li><li>Easy to apply</li></ul>	<ul><li>Need to ensure any excess is properly cleaned off as it will show as shiny marks if left</li></ul>

Cellulose paste modified with starch ether is probably the most popular choice for decorators for general hanging of papers, including lining papers and wood ingrain, because of its good adhesive and low marking properties.

Other pastes can be considered in certain circumstances, and the use of 'all-purpose' pastes has grown in popularity. These pastes provide all the benefits of starch and cellulose pastes. The thickness of the paste can easily be adjusted according to the type of paper being hung.

Adhesives can be supplied in powder form for self-mixing, or ready mixed in tubs. Tub-based pastes are often used with rollers when pasting.

Always follow correct procedures when disposing of any waste product to avoid contaminating the watercourse.

Preparing adhesives

When mixing pastes, particularly from powders or flakes, it is essential to get the consistency right and to avoid lumps. If the paste is lumpy, it will be difficult to flatten out when pasting and will be likely to show as lumps or blisters in the finished job. It is important to follow the manufacturer's instructions and adjust the consistency to suit the type of paper being used. Do not over-thin the paste, as this is likely to cause problems after pasting such as over-stretching, over-soaking and possible delamination of duplex papers.

If pastes are being stored overnight, make sure they are covered to avoid contamination from dust and debris that may settle on the top.

If pastes are kept for too long, they can become stale and unusable. Cellulose pastes in particular become thin, and starch pastes can become thin and start to smell.

INDUSTRY TIP
When mixing paste, it is best to create a swirling effect in the bucket of water and then sprinkle the dry adhesive into the centre while continuing to stir.

▲ Figure 5.35 Mixing paste

Adhesives defects

Some common defects associated with pasting wallpaper are described below.

Blisters

Blisters can be due to using adhesive that is too thin, which in turn leads to over-soaking. The paper will then blister when hung, as it is likely to be still expanding on the wall. Some of the blisters may remain after drying.

Delamination

Delamination of duplex or two-part papers is when the two layers separate. It can occur particularly if the paper has been allowed to over-soak or has been pasted with paste that is too thin. The paste applied will soak through the first layer, softening the adhesive used to laminate the two layers together when the paper was manufactured. Once this happens then the two layers can separate.

Stretching

This is usually caused by over-soaking, or by application of paste that is too thin. The paper continues to expand and if it is a pattern paper the pattern will not match up, as lengths will have stretched differently.

Contamination

Usually caused by poor pasting technique: either the paste has got onto the face of the paper from the surface of the paste table or the paste has wrapped around the edges of the paper during application. It is extremely important to keep the paste table clean at all times. It should be regularly wiped to ensure no paste is on the face of the table. Occasionally water stains on the existing surface can be softened by the new paste and this can lead to them staining through the newly applied paper. This can be remedied by ensuring the water stain is adequately sealed before applying new paper.

Open joints and loose edges

Uneven surfaces sometimes make it difficult for the lengths of paper to be properly butted together. If the paper is over-stretched to make them butt, sometimes the joints can spring back, leaving an open gap.

③ APPLY STANDARD PAPERS TO WALLS AND CEILINGS

There are a number of considerations when applying paper to ceilings and walls. The following section looks at the access equipment required as well as the tools and equipment you will use and the methods and techniques you will need to know about.

Preparation

To ensure good preparation of the underlying surface it is usual to apply a size coat before hanging foundation papers. Sizing will help the paper adhere better and will allow for good movement when hanging so that the paper slides into position and forms good butt joints. For more information on preparing surfaces for decoration, go to Chapter 3.

Access equipment

You are likely to need to apply wallpaper above ground level. A sound knowledge of the Work at Height Regulations 2005 (as amended) will enable you to correctly select the access equipment required for each task. The selection of equipment will often depend on space, height and the need to follow the correct procedures as defined in the risk assessment and method statement.

In many domestic interiors lack of space and the need for a low-risk approach will often dictate the use of hop-ups and lightweight staging. Steps and podium steps will be limited to small areas when working on ceilings but will be widely used for papering walls.

Lightweight platforms could be used with hop-ups or trestles to produce a platform of varying lengths. Handrails can be fitted depending on the set-up employed and if the risk assessment identifies them as a requirement.

▲ Figure 5.36 Trestle

Tower scaffolds may be chosen where space allows and the risk assessment requires there to be guard rails to prevent falling. They can be adapted to enable longer working platforms to be attached using towers at each end of the run. It may be possible to utilise a cut-down version of this structure for lower-height work.

Figure 5.37 shows a stair scaffold that features a walkthrough frame, enabling the user to gain access to the inside of the scaffold to go up and down the ladder when required. A stair scaffold should always be used for staircase work, to enable safe access to both ceiling and wall surfaces. They have guard rails and toe boards on the working platform to reduce the likelihood of falling. This equipment stands clear of the walls to allow the lengths of paper to be dropped through and finished off at the bottom of the wall.

Further information on equipment for working at height as well as regulations and safe practices can be found in Chapter 2.

Tools and equipment

The most commonly used tools for paperhanging are a paperhanging brush and shears and other trimming equipment such as retractable or snap-off blade trimming knives. Vinyl papers are often smoothed down using hard plastic wide spatulas to provide the required pressure for removing air pockets.

The tools and equipment described in the table below should be included in your tool kit.

▲ Figure 5.37 Stair scaffold

Equipment	Definition
Tape measure	• Used for measuring lengths and distances. It is a retractable metal tape, usually 5 m or 10 m long • It should be wiped clean with a little oil to ensure smooth running of the tape inside the case • Allow the tape measure to dry before storing
Folding rule	• Used for measuring lengths of wallpaper before cutting as well as measuring widths of cuts • A folding rule is typically 1 m long and folds into four to make it easy to store in the pocket of overalls • Wipe clean with a damp cloth or sponge and allow to dry before storing
Plumb bob	• Used to ensure that first and subsequent lengths of wallpaper are hung vertically • A small weight, usually made from steel, is suspended from a length of cord. Make sure it is still before checking for **plumb** • Keep the bob clean and untangle the line and wrap it around when not in use

Equipment	Definition
Chalk and line 	• Marking chalk lines is particularly useful for setting out the first length of paper to a ceiling • The image shows the self-chalking type of line, although it is possible to use a piece of string and chalk sticks. The line needs to be rubbed with chalk each time it is used • Ensure that chalks and lines are kept dry. If you are using the self-chalking type, the container will periodically require topping up with chalk
Paste brush 	• A paste brush is a 125 mm or 150 mm flat wall brush used to apply paste or adhesive to wall or surface coverings. It is also used to apply **size** to **absorbent** wall and ceiling surfaces before hanging papers • These brushes are usually made from bristle or synthetic fibre. Wash with warm soapy water to remove any adhesive, rinse thoroughly in clean water, then hang to dry in a well-ventilated room. If brushes are stored wet or damp, they can be prone to **mildew**

KEY TERMS

Size: a thin coat of glue or thinned paste applied to an absorbent surface before hanging wallpaper. Size helps to even out the absorbency so that papers do not stick too soon.

Absorbent: an absorbent surface soaks up liquids, for example bare plaster, bare timber. The more liquid a material can soak up, the more absorbent it is.

Mildew: sometimes referred to as mould, it is a fungus that produces a superficial growth on various kinds of damp surface. Mildew is typically seen as black spots that multiply.

Paste table 	• A paste table is used to lay out wallpapers for measuring, cutting, matching and pasting. It is usually made from wood or plastic and is typically 1.8 m long and 560 mm wide (about the width of lining papers), although this can vary between manufacturers • Paste tables are usually collapsible to make them easy to transport. Keep the face and edges of the board free of paste to avoid transfer to the face of the paper • Wipe down with warm soapy water when finished, rinse with clean water and allow to dry before storing in a dry, well-ventilated room
Pasting machine 	• A pasting machine can be used to apply paste to papers. This method can be extremely quick and ensures that the correct amount of paste is applied when correctly set up • The machine is filled with paste and the paper is pulled through rollers that apply the paste to the paper. When the correct length has been pulled through it is cut off ready for folding and soaking, before hanging

Equipment	Definition
Sponges	• Sponges are used for wiping excess paste from surfaces • They are usually made of synthetic material that allows water to be absorbed so that the wetness can be transferred to the surface in a reasonably controlled way • Wash sponges with warm soapy water and then rinse clean. Allow to dry thoroughly before storing in a well-ventilated area
Buckets	• Plastic buckets are commonly used for mixing paste or size and are then used as a container to hold the paste or size during application • They are also used to store water for use with a sponge to wipe paste off surfaces • Wash buckets clean with warm soapy water and then rinse clean before storing **INDUSTRY TIP** Stretch string or a rubber band across the middle of the top of the bucket when pasting. This gives you somewhere to rest the brush when not in use.
Spirit level	• A spirit level is used for checking the vertical positioning of wallpaper, as well as horizontal borders • Be careful when transferring levels around a room, as a spirit level is not always the most accurate method for this. It is generally not as accurate as a plumb bob, but is particularly useful for short lengths of paper, such as over doors or windows. A spirit level is only truly accurate over the length of the bubble in the middle • Wipe clean when finished before storing **INDUSTRY TIP** Avoid dropping a spirit level, as this can jolt the bubble and affect the levelling ability, making it inaccurate.
Laser level	• Using a laser level is an extremely accurate method of checking horizontal and vertical levels • A relatively low-cost level will be more than adequate for decorating purposes, as long as it has the capability for providing horizontal and vertical laser lines and can also tilt and turn

Equipment	Definition
Straight edge  **Straight edge with handle** 	• Straight edges are used by some decorators, particularly for trimming waste paper when up against straight edges such as skirtings, door frames and ceiling edges • The two types shown, with a handle and without, are typically 600 mm in length
Trimming knife with retractable blade  **Trimming knife with snap-off blade** 	• Some decorators like to use trimming knives when cutting around obstacles • It is best to learn the skills of using shears for cutting and trimming first, as there will be occasions when knives are not suitable and may tear the paper • Two types of trimming knife are illustrated, and in both cases extreme care must be taken when handling them
Sharps box 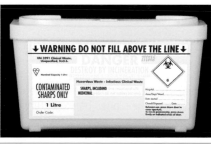	• When using the snap-off type of knife, there will be sharp edges that are snapped off and these should be immediately disposed of in a sharps box or container • When the container is nearly full, it should be taken to the local waste disposal area for correct disposal
Roller and scuttle 	• Some decorators like to use a roller and a scuttle filled with paste to speed up the process of pasting papers on the table • They are also useful for the types of papers that require the wall to be pasted instead of the back of the paper • You will still need to take care when applying paste to avoid getting it on the front of the paper

Equipment	Definition
Seam roller	• Used to roll down the edges of paper, or in corners or angles • Wipe the seam roller clean regularly when in use and wash it with warm soapy water when you have finished • Rinse clean and allow to dry – wait until it is fully dry before storing. A little oil may be applied to the spindle of the roller **INDUSTRY TIP** Use a piece of dry paper between the seam roller head and the wallpaper that is being rolled. This helps to avoid paste squeezing onto the surface of the roller and then being transferred to the face of the finished wallpaper and leaving marks.
Paperhanging brush	• Used to apply papers to walls to ensure that all air pockets are removed and that the paper lies flat without creases. It is sometimes referred to as a smoothing brush • Try to keep it clean and free of paste when in use • When you have finished, wash it in warm soapy water and rinse clean. Allow to dry thoroughly by hanging it up before storing flat in a well-ventilated place. As with paste brushes, if stored wet or damp these brushes can be susceptible to mildew
Paperhanging shears	• Used to cut lengths of wallpaper and for trimming around obstacles. Shears should be kept clean and sharp (they can be sharpened using a fine file or oilstone). Shears are also sometimes referred to as scissors • Shears can become clogged with paste and will require wiping clean to ensure that a good cut is maintained • Wash clean after use to remove paste, and store once dry to avoid the blades rusting. They can be lightly oiled around the pivot but ensure that any excess oil is removed from the blades before cutting wallpaper
Pencil	• A pencil is used primarily for marking out lengths of paper and for putting tick marks onto the wall to assist with marking plumb • Some decorators also use them for marking wallpapers at the top and bottom of hung lengths as well as around obstacles. • An HB pencil (not too hard) is ideal – make sure it always has a point • Marking for cutting can also be done with the back edge of the paperhanging shears. Using the back edge of the shears creates a crease line in the paper, thus avoiding the potential for leaving pencil marks

Equipment	Definition
Barrier cream 	• Thin latex protective gloves are especially useful if you are sensitive to the fungicide in some pastes. This chemical additive can cause irritation, and in some cases dermatitis • Barrier cream will provide some protection against irritation and should be used before carrying out any task • Always wash your hands with soap and water after papering tasks and before eating food to avoid ingesting any chemicals

General maintenance of tools

Ensure that all tools are cleaned after use and allowed to dry. Generally, all tools will need to be stored in dry, ventilated areas. Many can be kept in drawers, or on shelves or storage racks. It is important to keep all metal tools free from damp to avoid them becoming rusty, and ventilated to avoid mould or rust forming on items as they dry out. Paste brushes and paperhanging brushes should be hung up to dry and stored in dry conditions to avoid them being attacked by mildew. Be careful when storing sharp tools, particularly in the case of snap-off or retractable knives. Make sure that they are fully retracted before storing.

Though many paperhanging tasks will take place without the need for wearing a hard hat, safety boots or goggles, gloves and barrier cream may be necessary when handling some of the pastes containing fungicide. Overalls with a bib or a paperhanger's apron provide protection and the pockets enable you to carry a small number of tools such as a hanging brush and shears, making them readily available for use.

Take extreme care when handling paperhanging equipment, particularly items with sharp edges such as shears; you will need to be especially careful if they are the type with very sharp points.

▲ Figure 5.38 Apron with front pocket

▲ Figure 5.39 A painter's bib and brace overalls

Pasting, folding and application methods

Cutting and pasting methods

Perfecting the skill of using paperhanging shears is extremely important. This will require lots of practice to ensure that you cut straight and true each time without tearing the paper.

Shears can be used for all cutting activities, including cutting around obstacles. You will need to keep them sharp to cut accurately around obstacles.

Using trimming knives accurately and well will also require practice. They are particularly useful for cutting against edges and around obstacles. Safe use is always important when using sharp tools to ensure that you do not cut yourself. If you are using knives with retractable blades, make sure the blades are fully retracted when they are not in use – this is especially important if you store them in your paperhanging apron. When using trimming knives with snap-off blades you will need to take the same precautions, and in addition the snapped-off blades must be stored in a specially designed container.

▲ Figure 5.40 Snap-off trimming knife

One further tool that is sometimes used for cutting, particularly around obstacles, is the casing wheel. However, it has become less popular and is now very difficult to source. Casing wheels have largely been replaced by knives as they are difficult to keep sharp or re-sharpen. They are used in a similar way to a knife, but the wheel is rolled around the obstacle, cutting as it goes.

Pasting methods

There are a number of possible pasting methods, including pasting the paper on the bench using a brush or roller, pasting the wall using a brush or roller, applying paste using a pasting machine, or using a wallpaper pasting trough for ready-pasted papers. The pasting-the-wall technique is only suitable for those papers where the manufacturer has specified this method of application.

Pasting using a brush

This is a tried-and-tested method for pasting papers or walls. Care must be taken to ensure that the correct amount of paste is applied. Of all the methods, this is probably the slowest in terms of speed of application.

▲ Figure 5.41 Pasting using a brush

Pasting using a roller

Using a roller enables a speedier application of the paste. Again, care is needed to ensure that the correct amount of paste is applied, and no paste gets on the table or face of the paper.

Pasting using a pasting trough

This method is only suitable for ready-pasted papers. The wallpaper has a dried paste coating applied at the manufacturing stage, before sale. The cut length is rolled and immersed in a trough that contains water, allowed to soak and then pulled from the trough. Excess water is allowed to drain off before the paper is hung on the wall.

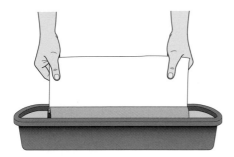

▲ Figure 5.42 Pasting trough

Pasting paper using a pasting machine

Paste is mixed and poured into the machine trough. The machine roller guides are set to the correct pressure and the wallpaper is pulled through, applying paste automatically to the face of the wallpaper. This is the quickest method of application.

Paste, fold and soak paper in line with manufacturer's instructions

Apply paste up through the middle of the length of paper and paste out towards the edge of the paste board, first towards the top edge and then pulling the paper across down towards the bottom edge (nearest you). Once the length of paper on the paste table has been pasted, begin folding the paper, ensuring that the edges are square when folded to stop them from drying out. Make sure that no paste gets onto the face of the paper or onto the paste table (from where it may be transferred to the face of the paper). Depending on the length of paper you may choose to use either the end-to-end folding technique or the concertina folding technique. Typically, end-to-end folding is selected for folding papers for standard height rooms when hanging them vertically. Concertina folding is used when hanging horizontally, for example when cross lining. This method is also used when hanging ceiling papers and can be employed for extremely long vertical lengths such as on staircases. It makes it easier to handle smaller folds when opening out and following the remainder of the hanging process.

Once the paper has been pasted, it should be allowed to soak for an appropriate length of time according to the manufacturer's instructions.

Under-soaking

If paper is not soaked enough this can lead to it having air bubbles underneath once hung – also known as **blistering**. This defect occurs because the paper is still trying to expand, and as it will be stuck in some places the paper will push outwards to form bubbles or blisters.

Over-soaking

If the paper has been allowed to soak for too long it will have over-stretched, and some of the paste may have started to dry. When paper over-soaks it will also be much harder to handle and will tear more easily. Duplex papers such as wood ingrain may delaminate when over-soaked. This is where the top layer separates from the bottom layer.

For more about these defects, see pages 225–226.

Folding

▲ Figure 5.43 Concertina fold

▲ Figure 5.44 End-to-end fold

Figure 5.43 and 5.44 show both concertina and end-to-end folding. Typically concertina folding is used for lining papers, ceiling papers and extra-long lengths such as those on a staircase.

End-to-end folding is used for most standard-sized rooms and makes it easier to mark papers that require cutting for width. This will be required when hanging to a corner, and each side of the corner will require cutting to width.

> **INDUSTRY TIP**
>
> Marking the top of the paper on the reverse with a pencil will ensure that you do not hang a piece upside down. This can easily happen with a plain paper.

Hanging techniques and processes

The processes described below will generally be appropriate to hanging foundation, plain and most pattern papers. Some papers, such as specialist types, will require special hanging instructions and some of these will be dealt with in the Level 3 specification.

Factors that need to be considered before beginning paperhanging include:

- calculating amount of paper required
- starting and finishing points relative to natural source of light or feature wall
- hanging method (horizontal or vertical)
- paper selection
- type of paste required
- pasting method to be used
- wall and ceiling features such as doors, features/obstacles, internal and external angles, sockets/switches/ceiling roses, borders and **window reveals**.

> **KEY TERM**
>
> **Window reveal:** the sides and head of the window, usually recessed in from the wall.

Try to ensure the room is cleared of furniture or that it has been arranged in the middle of the room

before you begin work. This is particularly important when papering the ceiling. Cover any items left in the room and lay a dust sheet in the area of the paste board.

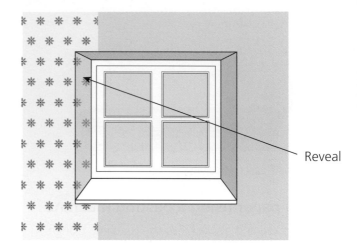
Reveal

▲ Figure 5.45 Window reveal

To ensure good preparation of the underlying surface it is usual to apply a size coat before hanging foundation papers. Sizing will help the papers to adhere better. It allows for good movement when hanging so that the papers slide into position, helping to provide good butt joints.

Papering walls
Starting and finishing points

The **starting point** in many rooms will be to hang the first length away from the light, i.e. a window or other source of daylight. Working away from the light in this manner will minimise shadows appearing along the edges of the joints and therefore they will be less noticeable.

If the room contains a feature such as a fireplace or feature wall, you should also consider this when planning how to achieve the best effect. Note the idea of **centring** over the fireplace to try to get an even pattern appearance, as shown in Figure 5.46.

> **KEY TERM**
>
> **Centring:** setting out a wall to create a balanced or even effect for the pattern. Working out from the centre should allow the pattern to be even in appearance.

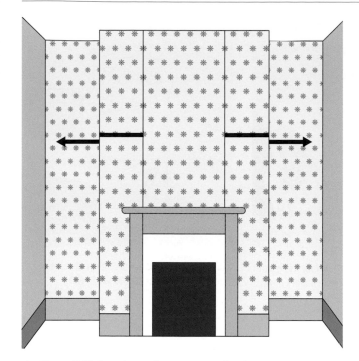

▲ Figure 5.46 Centring wallpaper over a fireplace

Centring over the centre line of the wall

In Figure 5.47 the pattern has been selected to maintain a strong design at the top of the paper and the wallpaper has been centred by hanging the first lengths *over* the centre line. The resulting appearance is balanced, and it should be noticed that at the sides of the wall, the wallpaper pattern and widths are equal in appearance.

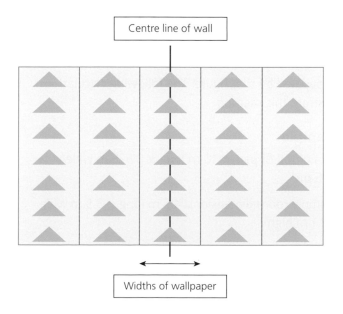

▲ Figure 5.47 Centring wallpaper over the centre line

Centring either side of the centre line

In Figure 5.48 the pattern has been selected to maintain a strong design at the top of the paper and the wallpaper has been centred by hanging the first lengths *either side* of the centre line. The resulting appearance is balanced, and it should be noticed that at the sides of the wall, the wallpaper pattern and widths are equal in appearance.

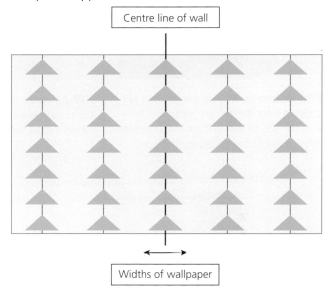

▲ Figure 5.48 Centring wallpaper either side of the centre line

The **finishing point** in many cases will be above the doorway entering the room. The main idea is to find an area where the loss of pattern will be less noticeable.

There will be a number of occasions when you will need to mark starting lines, including the first drop to be hung after internal and external angles. It is also necessary to provide a horizontally levelled line when cross lining. This line will be marked approximately one length down from the ceiling line to ensure that all following lengths remain horizontal. Cross lining is recommended when papering over excess filling or making good; over solvent-painted surfaces that offer little adhesion; and to even up the porosity of the surface before hanging the finishing paper. The lining paper should be hung horizontally to the vertical finishing paper.

Marking lines

Few walls are truly square or perfectly vertical. To overcome this, and avoid the pattern going out of alignment, it is essential to mark a vertical pencil line against a plumb line or long spirit level adjacent to where the first length is to hang.

Step-by-step marking lines

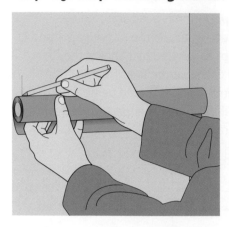

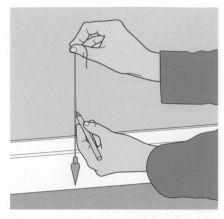

STEP 1 Measure the width of a roll or use a roll of paper to mark the wall, one width away from the corner (less 10 mm) to allow a slight overlap onto the return wall or chosen starting point.

STEP 2 Allow the plumb bob to swing freely until it is at rest, before putting a pencil mark down the wall behind the string.

STEP 3 Using a plumb line, lightly pencil some guide marks from ceiling to skirting.

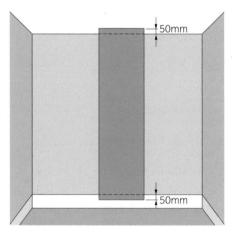

STEP 4 Measure your first length, allowing an extra 50 mm at the top and bottom.

STEP 5 Cut all lengths of paper on the table ready for pasting.

Measuring and calculating quantities of rolls of wallpaper

Before starting the paperhanging project, it is important to calculate how many rolls of paper will be required so that sufficient paper can be purchased before starting the job.

There are a few different methods used by decorators to calculate the quantity of paper required for a project, but the area and girthing methods are described here.

Area method

This method requires you to know the area of a roll of paper being used and to divide the total area to be papered by that figure. The following calculations illustrate this method.

Example

Area of roll of paper: 10 m long × 0.525 m wide = 5.25 m²

Example room dimensions: 4.5 m long × 3.5 m wide × 2.1 m high

You can calculate the area of the walls in a similar way.

To calculate the area of the room, add together the distance around the room (the **perimeter**) and then multiply by the height to give the total area.

In the example shown, the calculation will be as follows.

Area of the walls:

(4.5 m + 3.5 m + 4.5 m + 3.5 m) = 16 m (perimeter) × 2.1 m (height) = 33.6 m²

To work out how many rolls you need, divide the area of the walls by the area of a roll of paper:

33.6 m² ÷ 5.25 m² = 6.4 rolls of paper

This total should always be rounded up to the next full number, so in this case seven rolls will be required.

Note: This method of calculation will be perfectly adequate for calculating non-patterned wallpapers such as lining paper and wood ingrain, but for patterned papers more accurate methods should be used. The area method assumes that all the wallpaper can be used and makes little allowance for waste; it is also not as accurate as the girthing method

Girthing method

Another method of calculation is known as the girthing method. This is shown in Figure 5.49.

Using the measurements in the example room above, we will work out how many rolls of paper are needed.

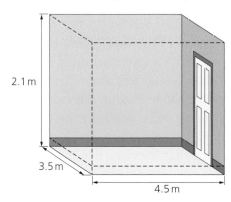

▲ Figure 5.49 Room dimensions

Example

In this example, we will assume that the rolls of paper have the following dimensions:

Roll length: 10 m

Roll width: 0.5 m (500 mm)

The room has the following dimensions:

4.5 m long × 3.5 m wide × 2.1 m high

The calculation is as follows:

Step 1: Add together all the sides of the room to calculate the perimeter:

4.5 + 3.5 + 4.5 + 3.5 = 16 m

Step 2: Divide the perimeter by the width of a roll of paper to calculate the number of lengths = 16 m ÷ 0.5 m = 32 lengths required

Step 3: You now need to know how many lengths can be cut from a roll. To do this, you need to know the room height, with 100 mm wastage for each length. This means that you are now measuring a room height of 2.2 m, rather than the original 2.1 m. Divide the length of one roll by the new room height to find out how many lengths can be cut from one roll.

10 ÷ 2.2 = 4.5

Round *down* to get the number of lengths per roll: 4.

Step 4: Divide the total number of lengths required by the number of lengths in one roll. In this example it will be:

32 ÷ 4 = **8 rolls**

A more practical way of following the girthing method is to mark the number of lengths around the room using a roll of wallpaper or a tape measure set at 525 mm, and to divide that total by 3. Three is the average number of lengths cut from one roll of wallpaper, particularly when using patterned wallpapers. You do need to consider the height of the room to be sure of the number that can be cut.

To make this a more accurate figure, the total area of items such as doors and windows needs to be deducted as well as an allowance made for pattern repeat.

In many cases, neither method shown will make sufficient allowance for waste, particularly when applied to drop patterns. For many domestic rooms of a standard 2.3 m height, it may only be possible to cut three lengths from a roll. In the example given this would mean buying an extra three rolls, as 11 rolls would be required. However, the girthing method is probably the more accurate of the two methods described here. As discussed earlier in this chapter, the additional waste caused by drop patterned papers can be minimised when cutting from two rolls.

Calculating quantities for ceilings

Quantities of paper for ceilings are normally calculated using a method that involves calculating how many lengths can be cut from a roll. Then divide the width of the room by the width of the roll of paper to work out how many lengths are required.

Example

Use the room sizes from the last illustration: 4.5 m long × 3.5 m wide × 2.1 m high

Length of roll of paper divided by length of room = 10 m ÷ 4.7 m (added 100 mm for cutting at each end) = 2.1 lengths, therefore this will be 2 lengths per roll.

Width of room = 3.5 m ÷ 0.5 m = 7 lengths required. Allow 8 lengths (best to allow extra length for any variations to room width due to room being out of square).

If 8 lengths are required and we can cut 2 lengths from a roll, then:

8 ÷ 2 = **4 rolls required**

IMPROVE YOUR MATHS

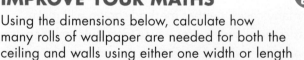

Using the dimensions below, calculate how many rolls of wallpaper are needed for both the ceiling and walls using either one width or length measurement. As in the previous examples, ignore any aspects of pattern or allowance for doors and windows.

Room size = 6.2 m long × 3 m wide × 2.3 m high.

(Answer provided on page 301.)

Hanging finishing papers

Cross lining

Sometimes it is necessary to apply lining paper to the surface before hanging finishing papers. This technique is known as cross lining and usually carried out to provide a smooth, even base for the finish paper. In cross lining the paper is hung horizontal to the finish paper as applying it in this way helps avoid prominent vertical seams and increases bonding strength. However, if the lining paper is to be painted it is usually hung vertically.

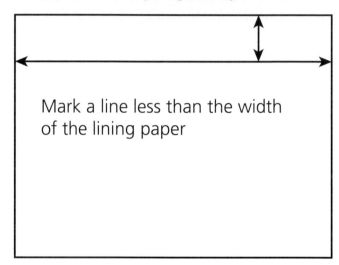

▲ Figure 5.50 Marking the first length for cross lining

Figure 5.50 shows the typical starting point for cross lining is at the top of the wall. Measure down from the ceiling and mark a level horizontal line on the wall with a chalk line at a position 50 mm less than the width of the paper. Making this mark less than the width of lining paper will allow for some excess to be cut along the ceiling line. This is done to allow for any variation in the ceiling line that can occur particularly in older properties. When hanging for cross lining ensure there are no overlaps and allow no more than a 1 mm gap at the joints. It is also advisable to cut the lining paper 1 mm short of obstacles and adjoining surfaces so that no lining is visible after the finishing paper has been hung.

Step-by-step process for papering walls

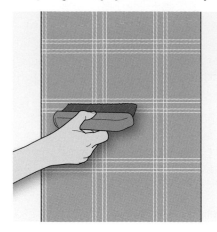

STEP 1 Having pasted the paper and allowed it to soak, in line with the manufacturer's instructions, hang the top fold against the plumbed line and brush out from the centre, working down.

STEP 2 When the paper is smoothly brushed down, run the outer edge of your scissors along the ceiling angle, peel away the paper, cut off the excess along the crease, then brush back onto the wall.

STEP 3 At the skirting, tap your brush gently into the top edge, peel away the paper and cut along the folded line with scissors or blade as before, then brush back.

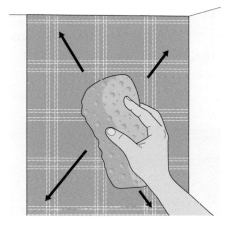

STEP 4 Using a sponge and a bucket of clean water, make sure you remove all of the paste from the surface of the paper.

STEP 5 Cut the next piece, allowing for pattern repeat, and paste and soak as before. Hang the piece, butting it up to the first, taking care to match the pattern. Do not overlap.

Papering around electrical fittings

Be very careful when applying wallpapers over, under or around electrical switches, sockets and similar items. It is recommended that you turn off the supply at the mains before carrying out this task, as it is possible to receive an electric shock as a result of the wet metal shears coming in contact with live areas.

- Smooth the wallpaper down very gently over the fitting and then, for square shapes, pierce the

paper in the centre to mark the corners and make diagonal cuts from the centre to each corner. Some decorators prefer to slightly unscrew the fitting so that the wallpaper can be tucked just behind.
- Press the wallpaper firmly around the edge of the fitting, lightly mark the outline and trim away the surplus.
- Press down around the outline, mark and trim in the same way.

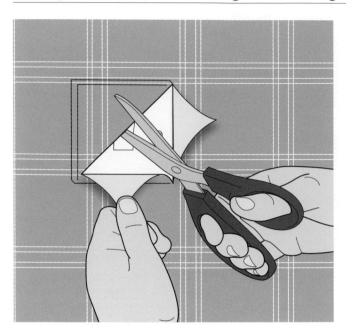

▲ Figure 5.51 Making diagonal cuts

▲ Figure 5.52 Trimming away the surplus

Papering behind radiators

Ideally, you should drain the radiator and take it off the wall so that you can paper behind it. If that is not possible, first turn off the heat and wait for the radiator to cool. Paste the strip of paper to the wall above the radiator, then slit it from the bottom edge so that you can smooth it down on either side of the radiator's fixing brackets. Press the paper in place behind the radiator, using a dry radiator paint roller.

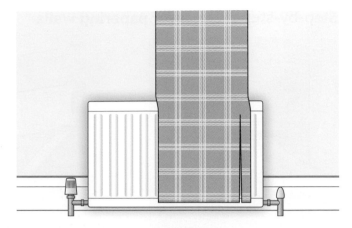

▲ Figure 5.53 Wallpapering behind a radiator

Papering around corners

Allow a minimum of 5 mm to turn into the corner. Turning as little as possible will minimise the amount of pattern lost. Sufficient turn should be allowed, however, for the corner to be possibly out of plumb.

When applying paper to corners, ensure that the return piece is plumbed so that the paper hung on the return wall will remain vertical.

▲ Figure 5.54 Wallpapering around a corner

Step-by-step papering around windows as the starting point

Decorators use a number of methods for papering around windows, and the following is a typical example used to get a good result.

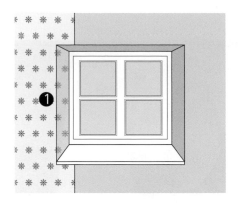

STEP 1 Hang the first length that goes into the recess and make cuts, top and bottom, to allow the spare piece to be brushed into the recess.

STEP 2 Join the next piece above the window and turn under the recess, then trim off the excess.

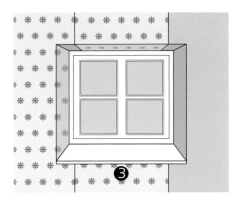

STEP 3 Join the next piece under the window and trim off the excess. Continue hanging the short lengths at the top and bottom of the window to fill spaces until the other side of the window is reached.

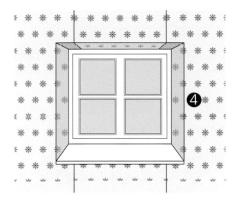

STEP 4 Hang the final length for the other side of the window as in Step 1. Ensure that this length is checked for plumb, as hanging around obstacles can set the paper out of plumb.

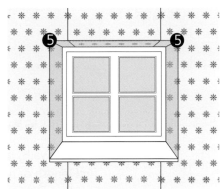

STEP 5 Cut and paste under the head of the window, slightly over-sized, to allow for laying under the edge of the paper already hung.

Papering ceilings

Papering ceilings involves some considerations that are different from when you paper walls.

Starting points

When papering ceilings, in most cases it is best to hang lengths working from the natural light source. This minimises the likelihood of any overlapped edges appearing when viewed from across the room.

When hanging patterned papers to ceilings, it may be desirable to centre the paper to enable an even pattern effect across the whole ceiling.

For odd-shaped or L-shaped rooms, it may be preferable to hang the longest length first and work away from there.

In some cases, economy of cutting and hanging will determine the direction of hanging. This should be considered as part of the planning process when you are calculating the quantity of paper required.

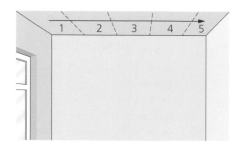

▲ Figure 5.55 Work away from the natural light source

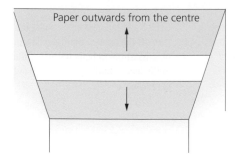

▲ Figure 5.56 Hang paper from the centre of the ceiling

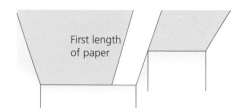

▲ Figure 5.57 Hang the longest piece of paper first in odd-shaped rooms

Marking out

▲ Figure 5.58 Mark out a starting point using a chalk line

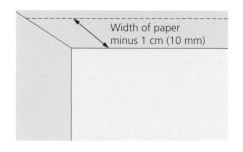

▲ Figure 5.59 Measure a width of paper out from the wall

As with walls, it is particularly important to get the first length of paper absolutely straight. Ceiling/wall junctions are not usually true enough to use as a guide. It is better to mark a chalk line and then position the first length along it.

Step-by-step instructions for papering ceilings

STEP 1 Paste can be applied to most papers by brush, roller or pasting machine. If the brush or roller method is employed, start pasting down through the centre, then outwards to the edges. Make sure no paste gets onto the edges of the paper.

STEP 2 After each paste board length has been pasted, start folding in a concertina fashion. Try to ensure that the folds are no greater than 350 mm, as if they are too wide this will make you need to stretch beyond arm's length.

STEP 3 Use a spare roll of paper or cardboard tube to act as a crutch to support the weight of the paper during the hanging process. Using this method will prevent paper from falling around your head as the length is opened.

STEP 4 Position the edge of the first length of paper against the chalk line and smooth the other edge into the ceiling/wall edge to give a 10 mm overlap onto the wall. If the edge of the wall is not straight, the overlap will be uneven, rather than the paper on the ceiling.

STEP 5 Brush out the bubbles with a paperhanging brush, using a sweeping action through the centre and brushing out to the edges. Continue opening each of the concertina folds while walking backwards along the scaffold. Always be aware of your footing and the edge of the scaffold platform to avoid stepping off the edge. If there is room to use a platform with guard rails, this will add extra protection.

STEP 6 Run wallpaper scissors along the ceiling/ wall edge to make a sharp crease. Gently pull back the paper and cut along the crease. Brush the trimmed edge back into place, applying extra paste at the edges if necessary. Butt the next length of paper against the first.

INDUSTRY TIP

Once the first length has been applied, check to ensure that enough paste is being transferred to the ceiling.

INDUSTRY TIP

It is dangerous to just drop waste offcuts onto the floor, as they are a potential slip or trip hazard.

Papering around electrical fittings

When papering ceilings it is quite likely that you will encounter obstacles such as ceiling roses – these require cutting around. It will take a good deal of practice to get this right while you are holding onto the full length of paper at the same time. The star cut technique should be employed. Find the centre of the rose with the point of the shears. Cuts can then be made from the centre outwards in a star fashion to the edge of the rose. Once this been done, push the rose through the paper, loosely press into position and then proceed with hanging the ceiling paper. Trim off around the rose using a trimming knife once the whole length has been properly smoothed into position.

INDUSTRY TIP

When carrying out this task, it is useful to remove the ceiling lamp holder cover, so switch off the power before you start. Ideally the complete obstacle would be removed altogether while papering ceilings, but this is definitely a job for an electrician.

Hanging wallpaper borders

When hanging wallpaper borders, it is usual to establish a level line using either a laser level or a long spirit level. These levels need to be accurately transferred around the room. Once the border has been cut for length and pasted using border adhesive, the border should be folded concertina-fashion. Hang the border in a similar way as you would hang paper on a ceiling, by opening folds as you go. Smooth out the folds with a smoothing brush, spatula or sponge.

If you are using border wallpaper to make frames or to emphasise doorways, it is almost certain that

▲ Figure 5.60 Cutting around a ceiling rose/lamp holder

you will want to make a mitre cut at each corner or junction. The illustration shows the border being cut at a 45° angle. Use a trimming knife to cut against a straight edge or wide-bladed scraper.

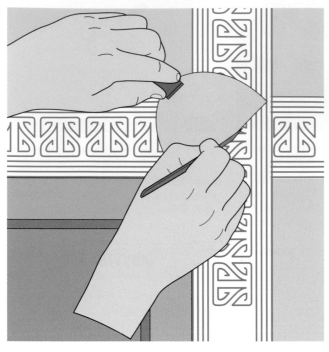

▲ Figure 5.61 Mitre cut in a corner

Papering staircases

Papering staircases presents slightly different challenges, although the actual hanging technique is the same. Not all staircases will follow the same process as shown in Figure 5.62, but this will give you an understanding of the principles to apply to the planning process.

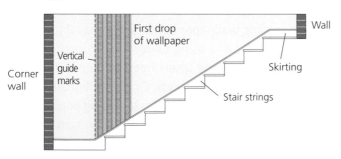

▲ Figure 5.62 Hanging paper on a staircase

The first length of paper to be hung should be as illustrated – by selecting the longest drop and plumbing a line. Hang the first length to this line and then work away in both directions until the wall is completed.

Access equipment for staircases

A stair scaffold with a walkthrough frame is ideal for working on staircases as it is adjustable to the slope of the stairs.

Good practice when papering

- Read the manufacturer's instructions first.
- Follow soaking times carefully – this will prevent shrinkage – and use the correct adhesive.
- Keep the face of the paper clean, removing adhesive with a sponge and clean water.
- Take care to apply an even amount of adhesive on the paper, particularly the edges.
- Make sure the wall is properly prepared, smooth and clean. Good preparation will ensure a better finish.
- Always use a sharp knife or scissors so as not to tear the paper.
- Double cutting or splicing is often a good way of achieving a perfect, invisible butt joint. This technique is particularly useful when using thick or heavily embossed papers (i.e. blown vinyl), where an overlap would clearly show. To achieve this, having overlapped the papers, use a sharp trimming knife and metal straight edge. Holding the straight edge down the middle of the overlap and pressing just hard enough to go through both layers of paper, draw the trimming knife from top to bottom. Remove the outer trimmed excess, then lift the top layer clear and remove the trimmed excess beneath. Brush the two layers flat to the wall. Wipe off the excess adhesive that will be on the face of one edge with a clean damp sponge.

Defects in paperhanging

Defects can occur due to a lack of care or following incorrect methods. Rectifying some of these defects can be very costly – often the defective wallpaper will need to be removed before hanging again.

Blistering

This is usually caused by under- or over-soaking and then careless brushing. Papers need to be properly pasted and smoothed (as described earlier in this chapter) to ensure that no air bubbles or excess paste are left to leave blisters on drying. Areas of paper that have been missed when pasting will be dry and will not stick to the wall surface, leading to blisters.

Creasing

This can be caused by excessive brushing which may cause the surface of the paper to stretch, then when trying to keep plumb there will be excess paper, which is likely to crease. Papering over very uneven surfaces can also cause paper to stretch, which will also result in excess paper and potential creasing.

Loss of emboss

Flattening of the edges or emboss of wallpaper will occur when poor brushing technique is employed to smooth out the paper. This can be made worse when the paper is left to over-soak or the paste has been applied too thin. Excess pressure applied in these cases will result in flattening.

Overlapping

This can be caused by over-brushing but is also likely to be due to over-soaking which can cause papers to stretch after hanging, leading to overlaps. Sometimes careless use of a seam roller can stretch the paper in this way.

Tearing

A range of things can cause paper to tear but in all cases it is likely to be due to a lack of care at some stage in the hanging process. If too thin paste is used this can cause over-wetting which in turn can make paper more likely to tear. Using blunt shears can cause paper to tear. Once papers have been soaked and are ready to hang, proper care should be taken as rough handling can lead papers to tear.

Springing of joints

This can occur when papers are over-stretched when butted together. They will shrink back on drying to leave an open joint. This can also be due to lack of adhesion and is particularly noticeable when papering on non-absorbent surfaces such as gloss painted walls.

Polished joints

In most cases this defect is caused by poor use of the seam roller. If paste is allowed to get onto the face of the paper at the edges, and is then rolled, it is likely that the seam roller will polish this area and leave shiny marks. This can be avoided by using a dry piece of paper between the seam roller and the face of the paper. Sometimes careless pasting can leave paste on the face of the paper edges and this also can give a polished effect when dried.

Shrinking

Excessive heat in a room will speed the drying of the pasted paper, potentially leading to shrinkage of the paper if it dries too quickly. This defect may also be caused by applying the paper to a porous surface, which allows the moisture from the paste to be absorbed rather than dry naturally.

Stretching

This defect is generally caused by over-soaking of the paper. The fibres in the paper absorb the paste and the paper will expand if left too long before hanging. Care should also be taken when hanging long lengths as the weight can sometimes cause the paper to over-stretch, particularly when applying wallpaper to staircases.

Open joints

Uneven surfaces can sometimes make it difficult for the lengths of paper to be properly butted together. If the paper is over-stretched to make them butt, then on occasion the joints can spring back, leaving an open gap.

Loose edges

This defect is almost always due to careless pasting techniques. Generally, these are edges that have been missed with paste or have an inadequate amount of paste covering. These dry areas will not stick to the surface and will lift up.

Irregular cutting

This is caused by poor technique when using cutting tools such as paperhanging shears. It is advisable to carry out a number of sessions of practice cutting of pasted paper before attempting to cut the real thing. The skill of cutting to a line without tearing the paper or producing irregular cuts is one that can be learnt with patience.

Inaccurate matching

This is usually due to poor application techniques but is made more difficult if the paper has been allowed to over-soak as the adjoining pieces may have soaked at different rates. Make sure the pattern is matched at eye level to minimise this defect.

Staining and surface marking

Usually caused by poor pasting technique, where either the paste has got onto the face from the paste table surface or the paste has been allowed to wrap around the edges when applying the paste. Always keep the paste table clean by wiping it regularly. Sometimes water stains on the existing surface can be softened by the new paste and can lead to them staining through the newly applied paper. This can be remedied by ensuring the water stain is adequately sealed before applying new paper.

Delamination

Delamination of duplex or two-part papers will often occur if the paper has been allowed to over-soak or has been pasted with too thin a paste. Delamination is also covered on page 204 of this chapter.

Corners incorrectly negotiated

It is extremely important to plan well when paperhanging, particularly when hanging pattern papers. With good planning it is possible to lessen the impact of pattern loss when turning corners and going around various obstacles. Starting points for paperhanging are very important, as are the correct application methods when dealing with corners (see page 220 for more on hanging in corners).

Inaccurate plumbing

Inaccurate plumbing will lead to lengths of paper being out of plumb, and on pattern papers this will be very noticeable as the pattern will start to run out of position at the top of the wall. The plumbing of the first lengths on turning corners must be carried out accurately and always be sure the plumb bob has stopped moving before marking. Hanging lengths of paper following obstacles like windows and doorways also requires the use of the plumb bob to establish a truly vertical starting point again.

> **ACTIVITY**
>
> Study the list of defects above and answer the following questions.
> 1 How are overlaps caused?
> 2 What causes flattening of embossed patterns?
> 3 What causes delamination of embossed papers?

Storage of paper and adhesives

When storing materials, consider their physical characteristics, bearing in mind that most will be badly affected by atmospheric conditions in terms of temperature, dampness or direct sunlight. Most of the products that you will be storing are made of paper or are dry, powder-based and sometimes contained in cardboard packets. Any opened products need to be properly sealed before being stored and you should also bear in mind that they have a limited shelf life before they become unusable.

Care should always be taken with the storage of wallpapers and adhesives, as they can be badly affected by damp, cold conditions. If it is too hot, papers can become too dry, and if they are exposed to direct sunlight discoloration will take place, which will particularly affect patterned papers.

Always refer to manufacturers' instructions about how to use, handle and store materials. However, in most cases they should be stored in warm, dry and secure conditions. If you buy materials in bulk, it is a good idea to keep a stock book to track when items are removed and require replacing. With all materials it is always best to check that oldest stock is used first to ensure it is not allowed to age too much. Most materials, particularly adhesives, will have use-by dates for guidance.

In the case of papers such as lining and ingrain papers, decorators will often store additional stock, as these products are used frequently – you will need a good racked storage system to do this. Wallpapers are best stored on racks and laid on their sides to prevent damage to their edges. Most papers are shrink wrapped in plastic to keep them clean before they are used. Paste and adhesives, similarly, should be stored in cool, dry, frost-free conditions where packets may be stored in drawers and tubs stored on shelves. If pastes are allowed to get damp they will be unusable, as they will start to become solid.

▲ Figure 5.63 Wallpaper is best stored in racks

▲ Figure 5.64 Pastes stored on shelves

ACTIVITY

1 What kind of storage conditions should be avoided?
2 What might happen to papers if rolls are stored on their ends?
3 Why are most papers shrink wrapped?

Safety considerations

Although you need to be aware of the requirements of the Health and Safety at Work Act 1974, particularly site rules about wearing personal protective equipment (PPE), it is probable that most paperhanging activities will take place in a closed room environment.

Other regulations that will need to be considered include the Control of Substances Hazardous to Health Regulations 2002 (COSHH), especially in relation to the handling of materials such as fungicidal paste. If any of the work is to be carried out at height you will need to follow the Work at Height Regulations 2005 (as amended).

Personal protective equipment

You will still need to comply with PPE requirements if you have to gain access across a site or within a refurbished property where wider site rules may apply. Remember that you will be required to wear a safety helmet, hi vis jacket, safety boots, gloves and goggles on some building sites as standard practice. However, it is usually acceptable to remove these while paperhanging. Do check this with your supervisor or site agent first, though.

Manual handling

Always follow the correct manual handling procedures when carrying out wallpapering tasks. Although in most cases the activity of paperhanging involves fairly light work, on occasion wall hangings can be supplied in particularly large and heavy boxes, and batches of wallpaper may be supplied in boxes of 12 rolls.

For more details, see Chapter 1.

Ventilation

Ensure there is good ventilation in the room when paperhanging as this will allow the pastes to dry out naturally and reduce any potential fumes that may be given off from the pastes or wallpapers.

Good ventilation will reduce the potential build-up of **volatile organic compounds** that may be contained in the wallpaper. Although VOCs are not generally present in large quantities, the wallpaper can have a strong odour when first hung.

KEY TERM

Volatile organic compounds (VOCs): emitted as gases from certain solids or liquids. VOCs include a variety of chemicals, some of which may have short- and long-term adverse health effects.

Avoid skin irritations

Many pastes and adhesives contain fungicide to reduce the likelihood of mould growth and this can irritate skin, and even lead to skin conditions such as dermatitis. If you have sensitive skin, wear gloves when decorating or wallpapering and use a barrier cream to help protect your skin. Always wash your hands, especially before eating food, to avoid absorbing or ingesting any product, as this could lead to stomach irritation or upset.

Product data sheets

Manufacturers have a legal duty under health and safety legislation to provide information about their products, and it is important to obtain a safety data sheet for each product used. Always follow the manufacturer's instructions and carry out a risk assessment to determine what the hazards are and how they can be reduced. The manufacturer's data sheet will include advice on first aid measures, such as how to deal with ingestion of the product, protective equipment required and disposal of waste products.

▲ Figure 5.65 First aid kit

▲ Figure 5.66 A safety data sheet for wallpaper adhesive

Waste disposal

Pastes are not generally described as hazardous, but waste materials should not be allowed to enter drains, soil or bodies of water. They should be collected and disposed of in the appropriate skip on site or in the correct area of the local authority recycling centre. Site rules regarding the need to keep potentially hazardous materials separate must be obeyed when disposing of waste products.

ACTIVITY

Obtain a wallpaper adhesive data sheet by finding one at your college or workplace or by searching online, and use it to help you answer the following questions.

1 What safety precautions should you take when mixing the adhesive?

2 Does the adhesive contain anything to inhibit mould growth?

Health and safety points to remember

- Pastes, paints and other materials can be harmful if you ingest them. Always wash your hands thoroughly, particularly before eating.
- Wallpaper adhesives contain a fungicide, so keep them away from animals and children.
- Do not smoke or eat when using wallpaper adhesives.
- When using sharp knives or scissors to trim wallpaper, cut away from yourself to avoid cuts or abrasions.
- If working around electrical switches or sockets make sure that the power is turned off at the mains.
- Do not stretch higher than you can reach – use a step ladder or hop-up and check they are set up correctly before use.
- Remember to wear the correct PPE and follow safe working practices when carrying out preparation processes. A face mask may be required when dry abrading areas that have been filled with powder-based fillers as there is a risk of inhaling excessive amounts of dust.

IMPROVE YOUR MATHS

Earlier in this chapter examples were given of how to calculate the correct amount of wallpaper when planning for a paperhanging project.

Using the girthing method, calculate how many rolls of paper are required to paper the walls of a room of the following size:

4.8 m long by 4.2 m wide and 2.3 m high

The rolls of paper have the following dimensions:

- Roll length: 10 m
- Roll width: 0.5 m (500 mm)

Make no allowance for pattern or other waste, and show all working out.

(Answer provided on page 301.)

▲ Figure 5.67 Using a proprietary wallpaper spatula to smooth vinyl wallpapers

Test your knowledge

1 By which manufacturing process is blown vinyl wallpaper produced?

  a Rotary printing

  b Wet embossing

  c Dry embossing

  d Heat expansion

2 By which process is Supaglypta produced?

  a Dry embossing

  b Wet embossing

  c Rotary printing

  d Heat expansion

3 Which type of wallpaper is produced using carved wooden blocks?

  a Machine print

  b Screen print

  c Hand block printing

  d Rotary printing

4 Which symbol relates to drop or offset pattern match?

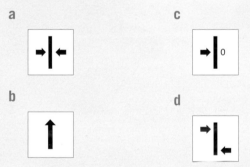

5 A paper known as duplex is made up of how many layers of paper?

  a 1 layer

  b 2 layers

  c 3 layers

  d 4 layers

6 Which wallpaper will provide a pattern in relief?

  a Pulp

  b Simplex

  c Embossed

  d Washable

7 What does reverse alternate lengths mean?

  a All lengths of paper hung upside down

  b Every second length is hung upside down

  c Every third length is hung upside down

  d Every fourth length is hung upside down

8 Which symbol indicates paper that has good light fastness?

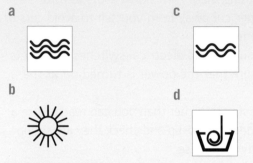

9 Which adhesive is most likely to encourage mould growth?

  a Border adhesive

  b Ready-mixed paste

  c Starch paste

  d Cellulose paste

10 Which application method will provide the quickest and most even application?

  a Applying paste to paper by pasting machine

  b Applying paste to paper by brush

  c Applying paste to paper by roller

  d Applying paste to paper by hand

11 Which defect could be caused by over-thinning paste?

  a Polished joints

  b Face staining

  c Springing joints

  d Over-stretching

12 Fungicides are used in some adhesives. What health problem can be caused by their use?

  a Difficulty breathing

  b Skin irritation

  c Dizzy spells

  d Loss of taste

13 Which paper is used as a lining for covering cracks and wood panelling?

a White lining paper

b Coloured lining paper

c Plain fibre lining

d Standard washable paper

14 Where is the suggested starting point for papering a ceiling?

a In a corner

b By the door

c By the window

d In the middle

15 What is the name given to checking for colour differences in rolls of wallpaper before hanging?

a Shading

b Blending

c Colouring

d Matching

16 Why should you vertically mark the starting point of the first length of paper?

a To make sure paper is upright

b To make sure you have enough paper

c To make sure you know where to finish

d To make sure you have the correct length

17 Where are star cuts used for cutting?

a Door openings

b Window openings

c Ceiling roses

d Chimney breasts

18 What is likely to have caused a loss of emboss?

a Over-brushing

b Poor matching

c Wrong batch

d Under-soaking

19 A dry room with a low level of light is ideal for storing which types of papers?

a Lining papers

b Pattern papers

c Wood ingrain

d Non-pattern papers

20 Why should rolls of wallpaper be stored in racks or bins?

a To make it easier to count rolls

b To prevent damage to the edges

c To see the pattern number

d To see the batch number

Practice assignment

Carry out the requirements of the synoptic assignment brief below in the practical workshop.

This section provides a short scenario-based activity that may provide preparation for the synoptic test.

Brief

A client has requested the redecoration of a Georgian-style bedroom, which contains a mixture of decorative features, including a panelled door and frame. The room also features a dado rail. The room has been previously painted with a white emulsion to the walls and is in good decorative order.

A basic layout of the door wall is provided in Figure 5.68. The wall has a height of 2.3 m.

The floor has recently been recarpeted, and the client has stressed that the carpet must not be damaged during the redecoration process.

The client has requested that the wall above the dado rail be cross lined with 1000 g lining paper and finished with a patterned wallpaper.

The client is looking for a professional job, to a high standard of finish, with no defects.

Task 1

Plan to carry out the work. You should produce the following:

● tool and equipment list

● method statement.

This should be completed before the practical work is attempted.

Task 2

Using Figure 5.68 as a guide, hang a minimum of five lengths of patterned wallpaper to demonstrate your skill of centring and to provide a well-balanced pattern. Note that the wallpaper will not be butted up to the door frame but should be planned to be cut around.

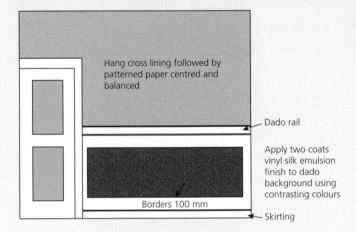

Hang cross lining followed by patterned paper centred and balanced

Dado rail

Apply two coats vinyl silk emulsion finish to dado background using contrasting colours

Borders 100 mm

Skirting

▲ Fgure 5.68 Layout of the door wall

CHAPTER 6

PRODUCING DECORATIVE EFFECTS

INTRODUCTION

This unit will provide you with the skills and knowledge required to create colour schemes and apply various decorative effects.

An understanding of colour theory is an important tool used by decorators to advise clients on potential colour schemes and provides a good basis for knowing how the use of colour in buildings can enrich our lives.

Decorative effects can enhance any decorating project. Their application requires creativity, an eye for detail and an artistic approach and they provide the professional decorator with traditional skills and knowledge that can be used in both the domestic and heritage sectors.

By the end of this chapter, you will have an understanding of:
- creating colour
- producing broken colour effects
- producing faux decorative effects.

The table below shows how the main headings in this chapter cover the learning outcomes for each qualification specification.

Chapter section	Level 1 Diploma in Painting and Decorating (6707-13) Unit 120	Level 2 Diploma in Painting and Decorating (6707-22/23) Unit 218	Level 2 Technical Certificate in Painting and Decorating (7907-20) Unit 206	Level 3 Advanced Technical Diploma in Painting and Decorating (7907-30) Unit 304	Level 2 City & Guilds NVQ Diploma in Decorative Finishing and Industrial Painting Occupations (6572-20) Unit 725
1. Understand creating colour	Not covered directly but an understanding of skills related to colour is developed at Level 2	Covered in Unit 230 1.1–1.8, 2.1–2.4, 3.1–3.9, 4.1–4.4, 5.1–5.3	1.1–1.2	N/A	This is an optional unit and is not a mandatory requirement for achieving the NVQ. The skills and knowledge acquired however will enable a decorator to work on a much broader range of contracts
2. Produce broken colour effects	1.1–1.5, 2.1–2.5, 3.1–3.14, 4.1–4.10	1.1–1.4, 2.1–2.8, 3.1–3.9	2.1–2.4	N/A	
3. Produce faux decorative effects	5.1–5.8, 6.1–6.8, 7.1–7.3	6.1–6.7, 7.1–7.7	3.1–3.3	2.1–2.3, 3.1–3.3	

1 UNDERSTAND CREATING COLOUR

▲ Figure 6.1 Selecting colours

As a decorator you have the privilege of regularly using colour in your day-to-day activities. Colour can influence mood, creating feelings of excitement or calm, and define space, form and texture. Developing a good understanding of the theory behind colour will allow you to advise clients on the types of decorative schemes that will enhance and suit their living space.

Theory of colour

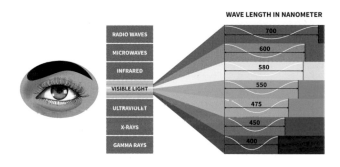

▲ Figure 6.2 Visible light is made up of different wavelengths

All colour originates in the way light behaves when it strikes a surface, so light is the source of all colour that we see. Light is made up of different wavelengths, and each wavelength is a particular colour. The colour we see is a result of the particular wavelengths that are reflected back to our eyes.

The way colour is seen differs for some people. For example, those with a perceived colour blindness have difficulty in seeing differences in colours. Red–green colour blindness is the most common form, followed by blue–yellow colour blindness and then

total colour blindness. Colour blindness is generally inherited and cannot be cured, and although this will have a minimal effect on your work as a decorator, it may have some bearing when selecting colour schemes. There is a specific test called an Ishihara test that can be carried out to determine whether colour blindness is present.

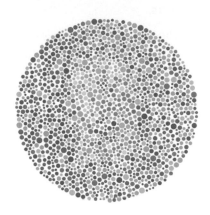

▲ Figure 6.3 Ishihara test letter B red and green

Pigments, dyes and other surfaces act as absorbers and reflectors of rays of light. For example, a red box absorbs most of the light falling on it, reflecting only rays from the red spectrum.

Pigment

A pigment is a material that changes the colour of reflected or transmitted light as a result of absorbing different wavelengths of light. Pigments are used for colouring paint and other materials.

The spectrum

The physicist Sir Isaac Newton made a scientific study of light rays and their relationship with colour. He discovered that sunlight is composed of all the colours in the **visible spectrum**. He used a triangular prism to bend the light rays, and after passing through the prism, the rays split into seven colours when projected onto white paper. You will have seen these colours in the form of a rainbow on a rainy day, which is formed by the sun's rays passing through the rain droplets.

Newton arranged this set of colours into a circular shape, which became the model for many future colour systems. It is said that the human eye can distinguish over 10 million different colours, but every colour is based on the colours of the light spectrum.

▲ Figure 6.4 A rainbow is formed by sunlight passing through rain droplets

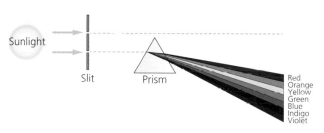

▲ Figure 6.5 Newton's theory of colour

Making colour

From prehistoric times colour has been used to provide decoration. The colour from plants and the earth allowed early cave dwellers to produce their art.

In more modern times, artists and decorators have continued to use natural products to make their paint colours. Paint manufacturers have adapted the ideas in order to mass produce paints, and still use naturally derived reds, blues and yellows along with manufactured pigments. Earth pigments such as ochre, umber, iron oxide and others are used to this day to provide some of the colours for paints. However, pigments are increasingly entirely artificially manufactured, with some using metals as their starting point and others produced from scratch by chemical processes.

▲ Figure 6.6 Cave dwellers used natural pigments to create colour paints

The colour wheel

In most colour wheels or circles the layout will be based on six of the seven spectrum colours, as follows: red, orange, yellow, green, blue and violet. Dark blue, or indigo (as this colour is sometimes called), is usually omitted in modern colour theory. The colour wheel can take various forms and may include reference numbers depending on what it is being used to show.

A colour wheel also helps us to develop colour schemes. The wheel allows us to see how colours work together and develop theories connected with them. The way we see colours and our **psychological** responses to them have inspired scientists and colourists to define terms to describe them and impose some kind of order on all the possibilities.

It is from the use of colour theory that paint manufacturers make sense of how colours go together. Later in this section you will look at how colour theory can be put to practical use when related to modern colour schemes and paint charts.

Primary colours

The simplest form of colour wheel is made up of only three colours – red, yellow and blue – and these are termed **primary** colours.

With the addition of black and white, the three primary colours can in theory be used to mix any colour. In traditional colour theory (as used in paint and pigments), the primary colours are the three pigment colours that cannot be formed or mixed by any combination of other colours.

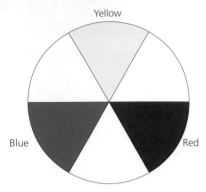

▲ Figure 6.7 Primary colours

Secondary colours

The second stage in building the colour wheel is to mix the primary colours with each other to form **secondary** colours.

- Red and yellow mixed together in equal amounts will produce the colour orange.
- Red and blue mixed together in equal amounts will produce the colour purple (or violet).
- Yellow and blue mixed together in equal amounts will produce the colour green.

If the colour mixes do not use equal amounts, the intensity of each of the colours will be altered.

KEY TERM

Secondary: second, or the second stage (after primary).

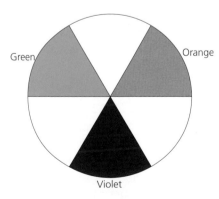

▲ Figure 6.8 Secondary colours

Tertiary colours

The third stage of extending the colour wheel is the development of a **tertiary** group of colours. A tertiary colour results from mixing a primary colour with an adjoining secondary colour, as shown in Figure 6.9. You can see why they are also referred to as intermediate colours: red added to orange makes red-orange, yellow added to orange makes yellow-orange, and so on.

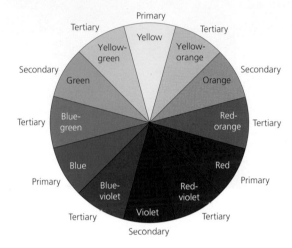

▲ Figure 6.9 Full circle showing primaries, secondaries and tertiary intermediaries

KEY TERM

Tertiary: third, or the third stage (after secondary).

Three more tertiary colours can be made by the admixture of the secondary colours with each other, as shown in Figure 6.10.

- Purple and orange mixed in equal amounts will produce a brown or russet colour.
- Orange and green mixed in equal amounts will produce an olive colour.
- Green and purple mixed in equal amounts will produce a slate colour.

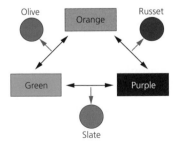

▲ Figure 6.10 Tertiary colours

Colour schemes

Using a colour wheel makes it possible to visually arrange colours to form schemes such as **analogous**, **complementary** and **monochromatic**. Understanding the ideas behind these schemes will allow you to apply them to interiors and exteriors for the benefit of your clients. In most cases you will find it possible to assemble colours that are pleasing to the eye and work well together.

Analogous colours

These are colours that are directly beside, or adjacent to, each other on the colour wheel. For example, a scheme using colours from the range of yellow, yellow-orange and orange would be described as analogous. One colour is usually used as a dominant colour, while others are used to enrich the scheme. The use of tints and shades of the pure **hues** will create a **tonal balance** and a harmonious overall effect.

KEY TERMS

Hue: a pure colour such as red or yellow.

Tonal balance: this is achieved by manipulating the use of colour. For example, a small amount of bright colour can offset the visual weight of a large area of less bright colour. Similarly, a small area of warm colour can balance a large area of cool colour.

ACTIVITY

State three other examples of analogous schemes from the illustration in Figure 6.9.

Complementary colours

Colours that are directly opposite each other on the colour wheel are known as complementary. Figure 6.11 shows that yellow and violet complement each other. Later in the chapter you will see how you can use this knowledge to good effect in colour schemes. Complementary colours are **contrasting** and give off a sense of energy, vigour and excitement. They will naturally go well together and one can be used to **accentuate** the other.

KEY WORDS

Contrast/contrasting: in colour terminology this usually relates to colours that are opposite on the colour wheel, which go well together.

Accent/accentuate: in colour terminology, using a small amount of contrast colour will enhance the other colour(s) and add excitement to a scheme.

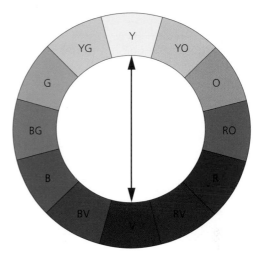

▲ Figure 6.11 Complementary colours

ACTIVITY

State three further examples of complementary colours.

Monochromatic colours

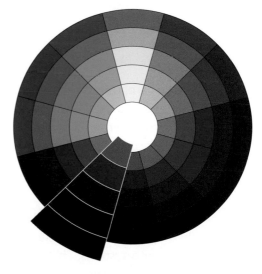

▲ Figure 6.12 Monochromatic colours

Figure 6.12 shows all the variations of one colour segment on the colour wheel, showing the dark, medium and light values of that colour. In this illustration, using the blue–violet hue at the outside of the circle, the term **monochromatic** is demonstrated by the addition of white. All or some of these colours, when used together, will provide a pleasing and **harmonious** scheme.

KEY TERMS

Monochromatic: all the colours (tones, tints and shades) of a single hue.

Harmony/harmonious: terms often used in the description of colour schemes to express that something is pleasing to look at because the colours look good together.

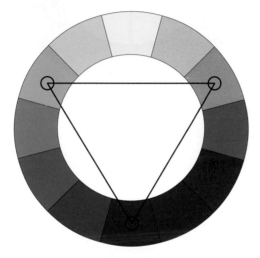

▲ Figure 6.13 Triadic colours

Triadic colours

A triadic colour scheme comprises three colours evenly spaced on the colour wheel. It is unlikely that they would be used in the same intensity and would most likely be used to provide accents of colour.

Colour organisational systems and terminology

The use of colour in the paint industry has led to the need to organise colour into systems. This helps with referencing and selection and ensures that colours conform to certain standards. Many paint scientists and colour theorists have also developed their own referencing systems.

Most of what we understand about colour was developed from the initial concepts of Sir Isaac Newton's interpretation of the colour spectrum (see page 234). It was not until the twentieth century that major advances in colour systems were proposed. Many of the tutors of the Bauhaus School of Design in Germany, such as Wassily Kandinsky and Johannes Itten, developed their concepts into something that could be studied and utilised in their artistic teachings. However, the following theories have primarily been further developed for use in our industry today.

The Munsell system

In 1915, Albert Munsell published the *Munsell Book of Colour*. He was an artist and an art teacher, and he developed the basic principles of his system mainly for the purpose of bringing order to the study of colour. Munsell's system was based on the three-dimensional attributes of **hue**, **value** and **chroma**.

Figure 6.14 shows how Munsell saw colour in a three-dimensional way. The Munsell colour system is set up as a numerical scale with visually uniform steps for each of the three colour attributes – in Munsell colour **notation**, each colour has a logical and visual relationship to all other colours. The central column is ranged white to dark at the bottom and the varying levels and leaves of the model depict the main colour gradually mixed with white, black and grey.

KEY TERMS

Value: in the Munsell system, colour's relative lightness or darkness.

Chroma: the degree of intensity, saturation, purity and brilliance of a colour.

Notation: text or numerical references that indicate the groups or categories of colours.

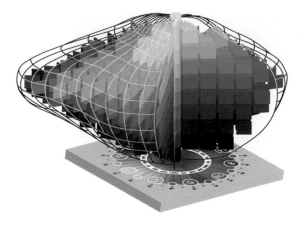

▲ Figure 6.14 Munsell solid

The Munsell system is based on the principles shown below, and these are also used by the BS 4800 British Standard for paint colour charts. In simple terms, the hue defines the colour against the value on the central column and its chroma will be determined by its relative closeness to the value column. The closer a colour is to the outside of the model, the more intense it will be.

- **Hue**: Each colour is described with a letter and there are ten principal hues. For example, 5Y represents a mid-yellow.
- **Value**: This represents the lightness (added white) or darkness (added black) of a colour and is identified by a number from 0 to 10.
- **Chroma**: This represents the greyness of a colour and is identified by a number from 0 to 14.

Specifying a colour using the Munsell system

A colour is fully specified by listing the three numbers for hue, value and chroma, in that order. For instance, a purple of medium lightness that is fairly saturated would be 5P 5/10, with 5P meaning the colour in the middle of the purple hue band, 5/ meaning medium value (lightness), and a chroma of 10, meaning it is quite intense (see Figure 6.15).

5P 5/10

▲ Figure 6.15 Example of a colour specified using the Munsell system (purple 5P 5/10)

In industry terms the number of colours that could be produced using this system was too great – it was felt that a smaller number of colours would more easily be accepted by manufacturers, as the cost of reproduction would be limited. Stockists would only need to stock a smaller range of colours and there would be fewer colours to show to clients when deciding on specifications. As a result, the British Standards Institute developed its own colour standards for the paint industry (see later in this chapter), although most of the systems in use are based on the Munsell system of hue, value and chroma.

Exploring value

Figure 6.16 shows how the value of a colour can be changed by the addition of the colour white. The squares within each column (1–9) have a red, green or blue hue, and in each column the same amount of white has been used, so the value of each of the column of colours is the same. This illustration demonstrates a range of colours that are monochromatic as well as demonstrating the term 'value'.

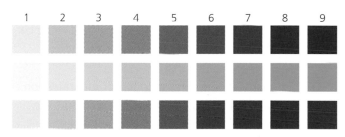

▲ Figure 6.16 The value of a colour is changed by the addition of white

ACTIVITY

Using watercolours or acrylic paints and a small brush, recreate the value image. With a pencil and ruler draw three rows of equal-sized squares onto cartridge paper and mark the squares from 1 to 9. Paint the 9 squares with the brightest red, green and blue, as in the image. Add small amounts of white each time to colours in successive squares until you reach square 1.

Industrial colour referencing

The major colour systems described below are the standards set by industry and/or the British Standards Institute.

BS 5252:1976 Framework for colour co-ordination for building purposes

BS 5252 is the British Standard that establishes a framework for co-ordinating the colours for all building products. There are 237 colours selected to enable standardisation between products such as paints, ceramics, plastics and tile flooring. It is now possible to obtain matches to any material using paint manufacturers' hand scanners. This range incorporates the BS 4800 colours but does not include the BS 381C standard range of colours.

Basic identification colours – BS 1710:1984			
12 D 45	Water	22 C 37	Acids or alkalis
10 A 03	Steam	20 E 51	Air
06 C 39	Oils	00 E 53	Other fluids
08 C 35	Gas	06 E 51	Electrical services

▲ Figure 6.17 Basic identification colours for pipelines and services (BS 1710)

INDUSTRY TIP

Paints, ceramics, tiles and plastics all have slightly different colour materials in their make-up. Therefore, it may not be possible to exactly match a paint colour to a bathroom suite using this system.

BS 4800

BS 4800 is a selection of colours for building purposes taken from the BS 5252 Framework for colour co-ordination. BS 4800 colours are widely specified in the UK.

BS 381C

BS 381C is a specification for colours for identification, coding and special purposes. These paint colours are technical colours used in industry and engineering including transport, and particularly by the Ministry of Defence for vehicles, buildings and signage.

BS 1710:1984 Specification for identification of pipelines and services

Colours from this standard are used to identify pipelines and services to enable engineers and other workers to recognise which pipes carry water, steam, gas and other chemicals or substances. Being able to identify services by colour standard allows engineers to carry out repairs safely. Figure 6.17 shows the colours used for pipeline identification according to this standard.

RAL colours

In 1927 the German Imperial Commission for Delivery Terms and Quality Assurance devised a colour range of 40 colours under the name RAL 480. In the 1930s the original notation system was changed to four digits and the collection was renamed RAL 840 R (R for revised). Colours were continually added to the range and it was revised again in 1961 with the RAL Classic range now consisting of 213 colours.

Pipe contents	RAL code
Gas	RAL 1004
Firefighting	RAL 3000
Air	RAL 5012
Combustible liquids	RAL 8001
Acids and alkalis	RAL 4001
Water	RAL 6010
Other liquids	RAL 9005
Steam	RAL 9006

▲ Figure 6.18 RAL colour codes for pipeline identification

RAL is the most popular European colour standard in use today. The colours are standard in architecture, construction, industry and road safety.

Figure 6.18 shows the colours used for pipeline identification using RAL four-digit numbers as a comparison with BS 1710, shown in Figure 6.17.

Natural colour system (NCS)

NCS is an international colour standard that is used to specify all types of surfaces. It is used by the majority of paint manufacturers and paint suppliers for their colour mixing machines to enable them to accurately match paints to a variety of materials.

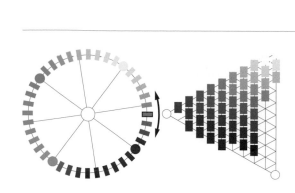

▲ Figure 6.19 The natural colour system colour wheel and a page of tints, shades and tones developed from the main hue

S 2570-Y60R

▲ Figure 6.20 The colour outlined in Figure 6.19 is shown here with its notation

INDUSTRY TIP

Each manufacturer will produce a range of colour cards or **swatches** that feature their most popular colours and finishes. This selection will usually be based on popularity and these cards are primarily targeted at clients to encourage them to select from that particular manufacturer's range.

KEY TERM

Swatch: a collection of paint, wallpaper or fabric samples, usually collected into a book form.

A manufacturer will tend to use its own referencing system for its colour range. A number of paint manufacturers now use hand scanners to try to match samples to their colour mix systems. It may therefore be possible to accurately match co-ordinating fabrics with paint colours.

For the purpose of this chapter we will explore BS 4800 further, as it is the industry standard colour system most used by the trade. However, many decorators also select from manufacturers' colour mixing systems, as they provide a much larger selection.

IMPROVE YOUR ENGLISH

Collect colour cards from three paint manufacturers for use later in this chapter. Briefly describe how each of them uses reference systems to organise the colour selection.

More about BS 4800:2011

BS 4800 specifies 122 colours of paint for building and construction work. This is an essential reference for anyone who needs a particular paint colour to use in the refurbishment of buildings – especially when working on local authority contracts or major works such as office blocks, airports, schools and hospitals. The original standard of 100 colours was updated to include 22 additional colours that are mostly brighter to reflect the latest trends for finishes on public buildings. These paint colours are widely recognised throughout the UK and are often used to meet safety, legal or contractual requirements.

The standard was primarily developed to create a system of colour notation ordered by numerical reference that the major paint manufacturers could follow. The individual colour notations enable colours to be mixed and sold to a standard, and each manufacturer follows the same formula for colour mixing. The BS 4800 colour system uses the Munsell principles of hue, value and chroma, but refers to them as hue, greyness and weight.

Figure 6.21 shows how the colours from BS 4800 may be arranged to form a colour wheel. It shows them in their brightest and most saturated form.

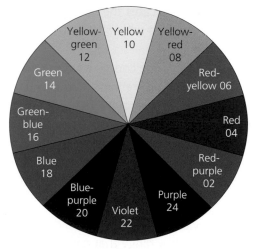

▲ Figure 6.21 12-colour wheel based on BS 4800

Figure 6.22 shows the greys plus the 12 hues used by BS 4800. Each colour selected will start with the hue reference number from this table. For example, a colour selected from the yellow range will begin with the number 10.

00 Grey plus black and white			
02 Red-purple		**14 Green**	
04 Red		**16 Green-blue**	
06 Red-yellow		**18 Blue**	
08 Yellow-red		**20 Blue-purple**	
10 Yellow		**22 Violet**	
12 Yellow-green		**24 Purple**	

▲ Figure 6.22 BS 4800 reference numbers

The colours used are for reference only, and manufacturers' colour cards should be used to define exact colours. Each BS 4800 colour for building purposes has a unique code – a combination of numbers and a letter, for example: 10 B 15.

This is what the code means:

- The first part of the code is an even number, from 00 to 24, which indicates the hue. In this case it is 10, and following the table above, this is from the yellow hue range.
- The second part of the code is a letter from A to E, and it indicates the greyness. Group A has a lot of grey content, while group E has little or no grey content. The letter here is B and the colour is in the yellow range (see Figure 6.23).
- The last part of the code is an odd number from 01 to 55. It indicates the weight – the higher the number, the brighter and stronger the colour.

This system provides a very precise way of describing a colour from the yellow range that contains a lot of grey and is not very intense.

This way of describing a colour means that it will be the same no matter which manufacturer produces it, even if they give it a more descriptive name such as 'Gardenia', 'Soft white' or 'Ivory'. The name may be different, but the code will always be the same, if it follows the British Standard.

In some cases, manufacturers will use the same name to describe a colour. For example, 'Magnolia' is a name often used, but unless the BS 4800 code is specified it is likely there will be some variation in colour.

ACTIVITY

Select a colour card from a paint manufacturer's range – preferably one that uses British Standard numbering. You may need to check with your tutor to make sure that the card uses this form of referencing. You will be able to use this colour card for other activities later in this chapter.

▲ Figure 6.23 10 B 15 is used on the walls in this room, making a fairly neutral colour scheme

ACTIVITY

Here are three BS 4800 codes:

- 10 E 55
- 00 A 13
- 04 D 44

Find the codes on your colour cards and describe them using the BS 4800 terms of hue, greyness and weight, in a similar way to the example given above.

Other colour terms

There are some other terms you may come across that are useful when describing colour.

Natural order of colour

The natural order starts with yellow being the lightest colour on the colour wheel (see Figure 6.9, page 236) and ends with purple as the darkest. Used in this order, the colours are pleasing to the eye and this helps us to make sense of how colour is used. A reversal of the natural order is known as **discord**. If colours are used in a reverse order, where the purple becomes the lightest colour and yellow the darkest, **discordant** colours are created.

Saturation

Many terms are used to describe **saturation** – this relates to the chroma, or 'weight', of the colour. The terms below are often used to denote the intensity, brightness or purity of a colour. In most cases, colours that appear fully saturated will seem to be very intense (strong) and will probably be used to provide accents. These fully saturated colours are often used for the front doors of properties, as they stand out from the (often) white painted frame.

Achromatic tones

Achromatic tones are not technically classed as colours, as they are without a hue. These range from black through to white and are sometimes referred to as sensations. The most popular achromatic colour scheme is black and white, often used in kitchens and bathrooms but shown in Figure 6.24 in a living room.

▲ Figure 6.24 Achromatic colour scheme

Neutrals

Neutrals are perhaps the easiest group of colours (or technically non-colours) to work with. They don't appear on the colour wheel and include black, grey, white and sometimes browns and beiges. They all go together and can be layered and mixed and matched because no neutral colour will dominate another. It is quite easy to add bolder colours to provide accents or excitement to the overall scheme.

Warm colours

Warm colours are usually associated with the yellow/red side of the colour wheel. They attract attention and are generally perceived as energetic or exciting. Warm colours are also sometimes referred to as **advancing colours** – they give the appearance of being closer to the eye.

▲ Figure 6.25 Warm colour scheme

Cool colours

These typically sit on the blue/green side of the colour wheel, opposite the warm colours, and they are generally perceived as soothing and calm. Cool colours are sometimes referred to as **receding**, as they create the appearance of space and appear to move further into the distance.

▲ Figure 6.26 Cool colour scheme

Tints and shades

A tint is a colour plus white. The addition of white makes the colour paler and less intense. A shade is a colour plus black. A dark blue shade, in other words, is blue that has been darkened by adding black.

Artificial light and colour

Colour can be greatly affected by both natural and artificial lighting, so when selecting colours, you should view them in the appropriate setting. Choosing paint colours in natural sunlight is ideal. Natural sunlight provides the neutral balance between the warm and cool ends of the light spectrum (with more yellow and blue, respectively). Northern light is the coolest, while southern exposure is the most intense. If you paint two rooms with the same paint – one with a northern exposure and one with a southern exposure – the wall colour will look different in each room.

Even natural sunlight isn't consistent. It changes throughout the day and varies depending on whether it is cloudy or clear. The shadows created by an overcast day will impact on how the wall colour looks, as well.

A bedroom that faces east and receives strong sunlight in the early morning will look very different when seen at night under artificial lighting. Most interior types of lighting cast a yellow aspect over colours and will alter their appearance accordingly. This should be taken into account to avoid disappointment about the appearance of a colour.

It is worth considering choosing a warm colour when decorating a room with a northern aspect, as it will not receive much sunlight and may therefore appear dark and cold.

Types of lighting

Different lighting types will have various effects and you need to consider this when selecting colours for interiors. This changing appearance of colour under different light sources is known as the **metameric effect**. If you try to match a colour under a particular lighting condition and then change the light source, the colour will generally no longer match.

▲ Figure 6.27 Metameric effect

Figure 6.27 shows how different colours appear when seen in fluorescent light and daylight. You can see from this that it is extremely important to ensure that when colours are selected they are viewed in the types of light to which they will be subjected. For example, tungsten and halogen lights enhance reds and yellows and mute blues and greens. Fluorescent lights enhance blues and greens and mute reds and yellows. To further complicate things, wall colour lit from above will look different from wall colour lit by floor and table lamps.

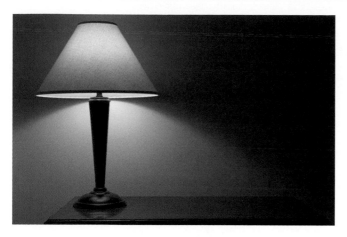

▲ Figure 6.28 Table lamps will make painted walls appear differently

▲ Figure 6.29 Fluorescent lighting provides an overall general brightness

▲ Figure 6.30 Sodium lighting makes the red building appear more brown in colour

The requirement to provide more **eco-friendly** types of lighting has led to the eco bulb, which tends to use a form of fluorescent, LED or halogen technology. These bulbs last longer than tungsten bulbs and are therefore a more sustainable option.

KEY TERM

Eco-friendly: products that consume less energy and are therefore less harmful to the environment.

Fluorescent lighting

This type of lighting has a green tinge, which can dull warm colours. Some give off a pink tinge, which is more colour-friendly and can even enhance warm colours. Fluorescent lighting tends to be used in offices or areas where general lighting is required. It provides an overall general brightness in an area but is not a substitute for natural light.

LPS (low pressure sodium) lighting

This kind of lighting is most often used for streetlamps and security lights. It gives off a soft, luminous glow with little glare. LPS is also sometimes used in cafes and restaurants, as it helps to create a warm, welcoming atmosphere. It is not typically used in domestic situations.

Halogen and LED (light-emitting diodes) lighting

These types of lighting are now commonly used in domestic situations, having migrated from more industrial or commercial settings. They are often used for accent lighting, spotlights and other general area lighting. **Halogen** and **LED** lights come closest to daylight in terms of how colours are affected – however, because bright light is cast on specific areas, there may well be areas of deep shadow where colours will look less bright. Halogen light bulbs are also now banned from sale under EU law although it is possible some households will still be using them.

In most cases it has become more environmentally sustainable to use LED lights as they are longer lasting and consume much less electricity.

KEY TERMS

Halogen: an incandescent lamp that has a small amount of halogen gas combined with a tungsten filament to produce a very bright white light.

LED (light-emitting diode): an electronic device that emits light when an electrical current is passed through it.

▲ Figure 6.31 The LED lighting used in this picture demonstrates how different areas can be accentuated

▲ Figure 6.33 Blue paint can produce a calming effect

Colour association

In very simple terms, colour association relates to how a person perceives various colours and how they affect the senses. For example, scientific tests have been carried out which show that red excites the senses and provokes feelings of warmth, while blue and green have been shown to invoke a sense of coolness. Purple is considered regal by some but may also be associated with death.

Psychological tests have also shown that there are some colour effects that have universal meaning. Colours in the red area of the colour spectrum are generally accepted as warm colours, though they encourage feelings of anger and hostility as well as warmth and comfort.

▲ Figure 6.32 Orange paint can provoke feelings of warmth

Colours on the blue side of the spectrum are generally seen as cool colours. These colours are often described as calm, but they also call to mind feelings of sadness.

As well as individual tastes, the perception of colour and colour preferences may be affected by cultural associations.

Colour schemes for internal and external areas

In this section, you will look at how to select colours and produce colour schemes for interior and external areas.

Paint manufacturers and various websites provide many ideas and choices when it comes to colour selection. However, if you apply the simple principles covered in this chapter you should be able to provide sound, accurate advice to your clients and hopefully meet their requirements.

Some manufacturers have developed 'heritage' or 'historical' ranges of colours specifically for organisations such as English Heritage and the National Trust. It has become increasingly desirable to be able to closely match Georgian, Victorian or other historical colours, particularly when recreating or refurbishing these types of property. Figure 6.34 shows a Georgian interior with typical use of calm green colours appropriate to the period.

▲ Figure 6.34 A Georgian-style room painted with heritage colours

Selecting colours for interiors

If your client has not selected colours for their interiors, you should start by considering the following key points:

- Which way does the room face – north, south, east or west?
- Is the room naturally light or dark?
- What is the room to be used for?
- Will the room be used mainly in artificial light?
- Are there items already in the scheme that could be used as a starting point?

▲ Figure 6.35 A strong feature wall

Other ideas that could be considered include the following:

- Provide a feature wall in a strong colour to contrast with neutral colours alongside.
- Paint out objects such as radiators in the same colour as the walls to make them less obvious.
- Look at colours from the swatch individually. Do not have them surrounded by lots of other colours from the colour swatch or chart.

▲ Figure 6.36 White radiators stand out. To make them less obvious, paint radiators the same shade as the wall

Creating a specification

Before starting, it is desirable to set out the scheme (e.g. monochrome, analogous or complementary) in the form of a specification or schedule, as shown in Figure 6.37. Using a table or system such as this will enable you to select your colour and also to put a colour chip representing that colour alongside the description. This will enhance communication with your client and minimise confusion.

Ceiling	The ceiling is to be prepared and finished with two fullcoats of vinyl silk emulsion in BS 08 C 31	
Wall filling	The wall filling is to be prepared and coated with BS 08 C 31	
Dado rail	To be prepared, primed with a water-borne primer/undercoat and brought to a two-coat acrylic gloss finish in BS 10 A 11	
Dado	The dado is to be prepared and coated with a ground colour of acrylic eggshell BS 10 A 03	
Skirting, door frames and architraves Window frames and architraves	To be prepared, primed with acrylic primer/undercoat and brought to a two-coat acrylic gloss finish in BS 10 A 11	

▲ Figure 6.37 Example specification

As part of the activities for this section, you will be required to illustrate monochromatic, analogous and complementary colour schemes and examples of each of these are shown below.

A monochromatic colour scheme, using green as the main hue is shown in Figure 6.38. A pastel green tint has been used on the back wall and stronger green colour has been used for the sofa to make a feature of it. Remember, monochromatic means bringing together colours using tints and shades of one hue.

▲ Figure 6.38 Monochromatic colour scheme

An analogous colour scheme, using red, yellow and orange in various tones is shown in Figure 6.39. Analogous means bringing together colours that are adjacent or side by side on the colour wheel.

▲ Figure 6.39 Analogous colour scheme

A complementary colour scheme, using predominantly orange and blue in various tones is shown in Figure 6.40. Complementary colour schemes bring together colours that are opposite each other on the colour wheel.

▲ Figure 6.40 Complementary colour scheme

ACTIVITY

Using the example specification shown in Figure 6.37, produce a colour scheme for each of the following:

- monochromatic
- analogous
- complementary.

Use colour cards to help you make your choices. As well as using colour names and reference numbers, stick the colour chip to the specification sheet. Make sure you clearly label each with the type of scheme, your name and the date.

Create a colour wheel on a broad surface

As part of your practical work, you may be required to set out, draw and apply the correct primary and secondary colours to a colour wheel on a broad surface such as a panel or wall.

Equipment

You will need the following equipment:

- pencils
- compass

- trammel
- ruler
- chalk line
- spirit level
- masking tape.

You will have come across most of this equipment before, except perhaps for the compass and trammel. These will allow you to draw circular shapes. The size of the circle will be limited only by how far the legs of the compass can be opened, or by how long the beam or trammel bar is.

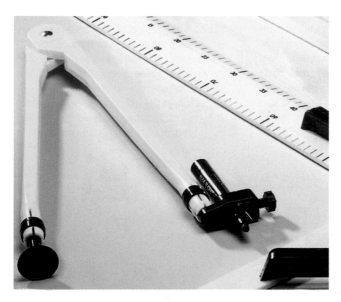

▲ Figure 6.41 Large board compass

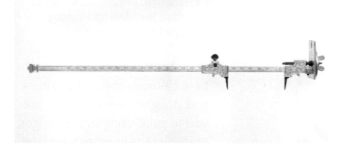

▲ Figure 6.42 Large beam compass or trammel

Draw a colour wheel to incorporate primary and secondary colours

Follow your tutor's instructions on the exact size and position of the wheel on the wall or board. You may use a large compass, trammel/beam compass or string and pencil.

INDUSTRY TIP

Stick a piece of putty or Blu Tack™ under the point of the compass to stop it slipping and also to protect the wall.

▲ Figure 6.43 Drawing a circle with a large compass

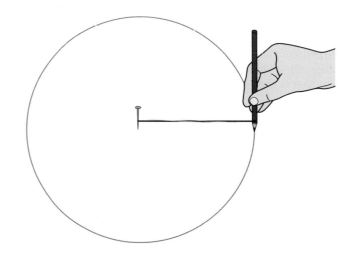

▲ Figure 6.44 Drawing a circle using a pencil and string

Step-by-step drawing a colour circle

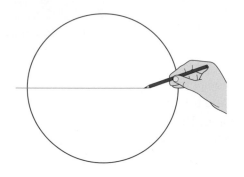

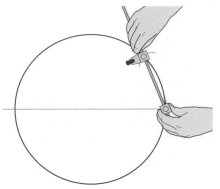

STEP 1 Once you have decided on the diameter and position of your circle, draw a circle using the board compass, trammel, or string and pencil.

STEP 2 Once the circle has been drawn to the correct size, draw a line horizontally through your centre point as in the illustration, using a spirit level to check that it is perfectly level. You now have two perfect halves to your circle and the line you have drawn is known as the diameter.

STEP 3 With your equipment still set to the same radius, place the point end on the outer edge of the diameter and make a mark on the edge of the circle with the pencil end. Move the compass point to that mark and make another mark further around the edge of the circle. Continue doing this until you have six marks, three to the top of the diameter and three to the bottom.

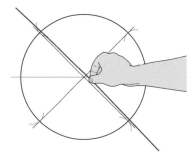

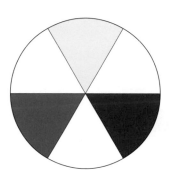

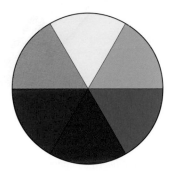

STEP 4 Using a line and chalk, join these marks diagonally as shown. Your drawn circle should now have six segments.

STEP 5 Apply the primary colours.

STEP 6 Using equal amounts, mix the correct primary colours to produce the secondary colours. Apply the secondary colours.

▲ Figure 6.45 A colour wheel can be set out using a trammel, board compass or string and pencil

Your tutors will advise you on the best methods of painting curves and straight lines. They may ask you to paint using straight edges, masking tape or other templates or freehand. Each method has its benefits, but with lots of practice you will be able to produce a clean, sharp colour wheel.

Case study

John and Razia are decorators and they have been asked to make recommendations for the colours at a house they are currently working on. The client, Maria, has asked for a colour scheme to go with a predominantly green carpet. The picture shows how it turned out. Using the client's choice of carpet colour has helped them to put together a very attractive scheme and the client was extremely pleased with the outcome. Using specific furnishing items to put a scheme together can be a very good starting point.

▲ Figure 6.46 Monochrome scheme using green

2 PRODUCE BROKEN COLOUR EFFECTS

This section of the chapter covers the skills and knowledge required to produce specialist decorative finishes, and you will draw on your experience of having prepared surfaces and painted them. You will develop your skills further by carrying out these decorative finishes, increasing your knowledge of a wide range of materials, tools and equipment.

Attention to detail, quality and cleanliness are most important, as this type of work is decorative and will be a focus for people to look at. Remember that not all painters can produce decorative finishes, and you will therefore be able to offer future employers and customers additional value with the skills you develop in this area of work. It will also enable you to increase your earnings, as you will be more skilled.

Personal protective equipment

You will need personal protective equipment (PPE) when you are carrying out any decorating work. The minimum precautions (safety measures) when producing specialist decorative finishes are as follows:

- Avoid products coming into contact with skin and eyes (wear gloves, and goggles if necessary).
- Ensure that there is good ventilation, using local exhaust ventilation (LEV) if necessary.
- Store and use products away from heat sources and flames.
- Do not eat or smoke in the vicinity of the work area.
- Wash hands before eating.

For more details see Chapter 1.

Produce quality finish ground coats for painted decorative work

▲ Figure 6.47 Sponge stippling by hand

It is possible to produce specialist finishes for decoration, and these can hide or mask minor surface imperfections, match other decoration and provide something different for the customer.

The advantages of these finishes are:
- each job is **unique**
- colours and effects can be tailored to suit the customer's personal choice.

The main disadvantages of specialist finishes are:
- it can be expensive to carry out this work, as it is labour intensive
- it is difficult, if not impossible, to repair damage to the finished work
- ideally, a smooth and level surface is required to work on.

Although the work of preparing and **grounding out** the surface may seem to be the least productive or enjoyable aspect of painting and decorating, it is important to carry it out thoroughly. Any imperfections may be made more obvious by the materials used for the decorative finish.

▲ Figure 6.48 Thorough preparation is important to avoid defects in the finish

Surfaces

The specialist finishes for which you will be developing skills are usually applied to walls, doors (e.g. kitchen cupboards or wardrobes) or items of furniture (e.g. boxes, picture/mirror frames). The substrates will therefore normally be previously painted timber, previously painted plaster or previously painted plasterboard.

Ground coats

The paint applied to the surface on which decorative finishes will be produced is called the **ground coat**, and it is important that it has a quality finish with no defects or irregularities, no excessive brush or roller marks and good **opacity**. It should show through the broken colour of the **scumble** and add depth to the surface by being part of the colour scheme.

It is important therefore to be aware of the overall colour scheme and where the decorative effect will be applied, to make sure that the ground coat and scumble colours complement or match the surrounding area.

Water-borne or solvent-borne paint systems with an eggshell or low-sheen finish are suitable for the ground coat, as matt or gloss finishes will adversely affect the **manipulation** process of the scumble for broken colour work. A matt finish will tend to absorb the scumble, while the manipulation tools will tend to skid around on a gloss finish. Pay attention to the compatibility of the ground coat material and the scumble type (i.e. how well they can be used together) and think about the location of the decorative effects in terms of durability. The colour relationship between the ground coat and the scumble is important – they should not be too different, to avoid a harsh finished appearance.

▲ Figure 6.49 Colour charts can be used to check that colours complement one another

Sample boards

It is advisable to produce sample boards before carrying out the work, so the customer can see a variety of colour and effect combinations and get an idea of what their chosen decorative scheme will look like. This may save you time and money in the long run, as it reduces the risk of the customer changing their mind about the work that has been produced or complaining that the end product is not what they had agreed to. Make sure that you accurately record on the reverse of each sample all the materials, quantities and proportions as well as the tools used, so that you can **replicate** the effect. Ask the customer to sign the back of the sample board to confirm their choice.

Preparation processes

To achieve the required quality finish for the ground coat, you need to abrade the surface thoroughly. Dry abrading may be appropriate as the first stage for poor, rough surfaces, but this should be followed by wet abrading (or wet flatting) to improve the finish by completely removing any application defects such as **ropiness**, runs or sags. This will produce a finish that is very smooth but keyed to ensure that the ground coat adheres.

It may be necessary to **make good** any imperfections or indentations (see Chapter 3, page 140) to ensure that the surface is level. The materials used for this will depend on the surface type and imperfection, including its depth. Repairs to timber may require a stopper; powder or ready-mixed filler will be appropriate for plaster or plasterboard, while shallow indentations will require fine surface filler.

It is important to select the correct abrasive type and grade for the stage of preparation. Use a coarser grade (e.g. 120) at first and finish with a fine grade (e.g. 400). The abrasive type must also be suitable for the dry method (e.g. aluminium oxide) or the wet method (e.g. silicon carbide) of abrading.

KEY TERMS

Ropiness: a surface finish defect similar to brush marks, but where the marks are much heavier and coarser; being more pronounced, they are highly visible and unsightly.

Making good: checking a surface for defects such as holes and cracks and applying fillers to make the surface smooth and without defect.

Spot-primed: to apply appropriate primer to sections of surface area that have been made good, to prevent the next coat from sinking into the filler.

Sinking: reduction in the sheen of a paint film. This may occur when a section of making good has not been spot-primed and the film former has been partly absorbed by the porous filler.

De-nibbed: the process of lightly sanding the surface to remove bits and nibs of dust.

▲ Figure 6.50 Make good indentations by using filler

ACTIVITY

Check the quality of the finished ground coat using your fingertips and by examining the surface from different angles to ensure it is of a high standard.

Areas that have been made good should be **spot-primed** to prevent the ground coat **sinking**. After the initial preparation, each coat of paint should be **de-nibbed** to remove any bittiness that may be present, followed by using a tack rag to wipe the surface and ensure that there is no residue or dust. For more information on preparation processes, see Chapter 3.

Tools and equipment required for the preparation process

You will need a variety of tools and equipment for the preparation process. The table that follows explains what these tools will be used for.

Tools and equipment	Description and uses
Rubbing blocks 	• Rubbing blocks hold abrasive paper and ensure that a flat, smooth finish is achieved • Rigid blocks may be made from rubber or cork • Flexible sanding sponges are suited to complex or curved surfaces, although the most hard-wearing type is rubber
Sponges 	• Used to wet the surface and then wash and wipe off the residue when using the wet abrading process • Choose a synthetic decorator's sponge, as opposed to the type that is used for washing cars

Tools and equipment	Description and uses
Buckets	• Used to hold water for the wet abrading process • Usually made of plastic
Dusting brush	• Used to remove dust and dirt from the surface and surrounding area before painting
Work area protection	• Dust sheets protect floors • Masking paper and masking tape may also be required to protect adjacent surfaces

Personal protective equipment

The personal protective equipment (PPE) required for the preparation process is described in the table in Chapter 1 (pages 30–31).

The painting process

Thorough preparation must be followed by careful paint application, using clean paint, clean application tools and a clean paint kettle or roller tray. This can be achieved by:

- dusting off the lid of the container before opening
- carefully removing and disposing of any skin that may have formed on the paint surface
- stirring the paint
- straining the paint if it was taken from a previously used container
- adjusting the consistency of the paint to take account of atmospheric conditions and/or surface

conditions, by using the appropriate thinner (water for water-borne paint and white spirit for solvent-borne paint).

ACTIVITY

Using the internet, find information about how atmospheric conditions can affect water-borne coatings.

Number of coats

The number of coats of paint required will depend on whether:

- priming is required (due to the method and extent of preparation of the surface)
- there will be a strong colour change from a previous coating

- the colour has poor opacity, for example certain yellows or blues
- the coating type is changing, for example from solvent-borne to water-borne.

Tools and equipment required for the painting process

Using the right tools and equipment is key to producing clean finishes.

Paint stirrers

Paint stirrers are used to ensure that all the ingredients in the container are dispersed evenly and that the coating is of a smooth consistency. Make sure you use clean, dry, smooth stirrers so they do not introduce contamination.

▲ Figure 6.51 Paint stirrers

Strainers

Strainers should be used for coatings taken from a previously used container to remove any contamination such as dirt or paint skin. If you are combining colours, a strainer also reduces the chance of there being any unmixed pigment. **Proprietary strainers** are most appropriate for larger quantities of paint or scumble.

> **KEY TERM**
>
> **Proprietary strainers:** manufacturers' ready-made strainers provide an open mesh that filters out the bits, leaving clean material that is free from contamination.

▲ Figure 6.52 Paint strainer

> **INDUSTRY TIP**
>
> Fine mesh stockings or tights are fine for straining smaller quantities and are economical and disposable.

Paint brushes

Natural bristle brushes should be used for solvent-borne coatings, and synthetic filament brushes for water-borne coatings. There are a range of sizes to choose from that can be used according to the surface area to be coated.

▲ Figure 6.53 Select the right brush for the coating you are using

Rollers

These should have a mohair sleeve, so no texture is left in the wet film; a range of sizes is available depending on the surface area to be coated.

▲ Figure 6.54 A mohair roller sleeve

Kettles and roller trays

Use galvanised (metal) kettles for solvent-borne coatings, and plastic pots for water-borne coatings. The size of roller trays should be appropriate for the rollers being used.

▲ Figure 6.55 Use metal kettles for solvent-borne coatings

Hair stipple brushes

These are often abbreviated to hair stipplers or **stipplers** and are used to remove all traces of brush marks and leave a smooth, even finish. These brushes come in different sizes (dimensions are given in inches), and particular care should be taken when using, cleaning and storing them. When not in use, stipplers should be laid on their sides rather than left standing on the bristle ends, as this may damage and distort the bristles. The effectiveness of the brush depends on the broad, flat area of bristle tips. Avoid build-up of paint or scumble on the bristles by frequently wiping the bristle ends with a cloth dampened with the appropriate thinner.

▲ Figure 6.56 Hair stipple brush – 6 × 4

▲ Figure 6.57 Hair stipple brush – 4 × 1

There are also rubber stipple brushes available, which are normally used for texture paint. If a bold effect is required, or the effect will be viewed from a distance, these may be used to manipulate the scumble. It is best to use this type of stippler with water-borne products, as cleaning solvents will cause the brush to deteriorate.

To maintain the condition of the brush, clean it as soon as it is no longer needed, to avoid the coating drying in the bristles. Place two or three trays of thinner with a pile of rags between them. Place the brush in each tray in turn, using the rags to absorb excess thinner between trays as the brush gets gradually cleaner. When all traces of paint or scumble have been removed, wash the brush in warm, soapy water and rinse it to remove any final traces of thinner and keep the bristles soft. Hang it up to prevent damage to the bristles and make sure it is fully dry before storing it, preferably in a box to protect it.

ACTIVITY

Using your training centre's resources, find out the source of 'natural bristles', the main country of origin and supply problems (if any).

Tack rags/cloths

These are used just before applying coatings to pick up loose particles of dust, dirt and lint from the surface. Lint-free tack cloths are made from a continuous (non-fibrous) filament of synthetic yarn. They have finished edges and have been coated in a non-drying resin. When not in use, the tack rag or cloth should be stored in its polythene wrapping. Keep it clean and away from direct sunlight to prevent the resin drying.

Defects

As the materials used to produce broken colour effects are generally quite thin and semi-transparent, defects will be evident when:

- there is insufficient surface preparation
- dirty or contaminated paint, application tools and equipment have been used
- inappropriate application methods have been used.

Bittiness in the ground coat is unsightly and the scumble can collect around this, making it appear a darker colour and spoiling the finished effect.

Heavy-handed laying off by brush will result in brush marks or ropiness and, as with bittiness, the scumble can collect in the uneven paint film. Careless laying off may also lead to misses in the ground coat, meaning the scumble will sink.

Poor opacity or inconsistent laying off can also leave an uneven ground coat colour which may be visible in the finished work.

▲ Figure 6.58 An example of bittiness in a ground coat

To help reduce or avoid application defects in the ground coat, it is best to use either a roller with a mohair sleeve or a hair stipple brush to provide a finish that is free from brush marks.

INDUSTRY TIP

For speed and evenness of application, use a roller to apply the ground coat. This will reduce the amount of work required to produce an even coating with a hair stipple brush.

Environmental and health and safety considerations

Attention to safe and healthy working practices is important because you, other people and the environment are all affected by your work activities. Legislation is in place to help protect all three. Environmental and health and safety regulations are covered in more detail in Chapter 1, but it is important to be aware of how they relate to producing specialist finishes.

ACTIVITY

Find out what the word 'legislation' means, and who it will affect in construction.

Control of Substances Hazardous to Health (COSHH) Regulations 2002

These regulations are intended to protect employees and others from the effects of working with substances that are hazardous to health. The regulations cover substances that:

- produce gas, fumes, dust, mist or vapour which may be breathed in (inhaled)
- damage or contaminate skin
- could be swallowed by being transferred from hands to mouth
- irritate eyes or could permanently damage eyesight
- may enter the body through a skin puncture (these are quite rare).

When working to produce your ground coat, you may be generating:

- dust, from the dry abrading process
- residue, from the wet abrading process – this could contain lead in an older (pre-1970) building
- fumes or volatile organic compounds (VOCs) when applying paints. Remember that, contrary to popular belief, water-borne coatings are not solvent free.

▲ Figure 6.59 COSHH symbol for irritant

In addition to the use of dust masks, disposable gloves and barrier cream, all of which will help to protect you while you carry out decorating activities, you also need a good level of ventilation in the work area to ensure that there is no build-up of fumes or dust. All of the above are together known as control measures.

You must make use of any control measures and equipment provided and comply with any special arrangements the employer (or your place of learning) has put into effect.

Work at Height Regulations 2005 (as amended)

A place is considered 'at height' if a person could be injured falling from it, even if it is at, or below, ground level. The Work at Height Regulations are designed to prevent the deaths and injuries caused each year by falls at work.

Under this legislation, you are required to do all that is **reasonably practicable** to prevent anyone falling. For further information about the Work at Height Regulations 2005 (as amended), see Chapters 1 and 2.

While producing specialist finishes, you may be required to work from stepladders or podiums/hop-ups. Chapter 2 deals with various types of access equipment in detail. Always check equipment for defects before use, and make sure it is secure and on firm, level ground before starting to use it.

▲ Figure 6.60 Working at height is any work where there is the risk of falling

Electrical safety

If you use an electrical sander to prepare the surface, use it in conjunction with a 110 V transformer to reduce the risk of death by electrocution. Check all plugs, cables and connectors before use. They should be in good condition – if there is any sign of damage, this should be reported and the equipment must not be used.

▲ Figure 6.61 Check electrical equipment before use

Disposal of waste

As with health and safety, there is legislation to control the safe disposal of waste products, including coatings that are left over, used thinners, used rags and so on. The Environment Agency is responsible for regulating the disposal of waste, and you should ensure that you comply with the regulations governing different kinds of waste.

Your local authority will normally provide waste disposal services and can give you guidance on how to dispose of your particular waste. There may be a charge for this, as it is seen as a commercial side of their business. You may therefore need to build the cost for waste disposal into the price you quote for any job.

ACTIVITY

Select three materials you might use for decorative finishing work. Visit the Environment Agency website (www.gov.uk/government/organisations/environment-agency) to find the suggested safe disposal method for each one.

Risk assessments

A risk assessment will be needed in order to work and use materials in a safe and healthy way. As explained in previous chapters this is a simple look at whether any aspects of the planned work could cause harm to people. You can then decide whether you have taken enough precautions, or whether you need to do more to prevent harm.

ACTIVITY

Undertake the first step of a risk assessment by listing all the hazardous activities and materials involved in the preparation and painting of a surface to produce a high-quality ground coat.

Produce broken colour effects using acrylic and oil-based scumbles

This section is about producing broken colour effects by sponge stippling and two methods of **rag rolling**. A translucent coloured glaze called scumble is applied over a different-coloured ground coat to create the desired effect. The translucency of the scumble will allow the ground coat to be seen through it and will be part of the colour scheme.

KEY TERM

Rag rolling: involves the rolling of lint-free rags or a chamois leather into a rough ball and rolled around in a random fashion over the surface. The coloured glaze is manipulated and takes the form of crushed velvet. Avoid using in straight lines as a banded appearance is likely to be formed.

The two methods of rag rolling are as follows:
- **Subtractive method**: Scumble is applied in a continuous, even film to the ground coat, then manipulated using crumpled lint-free fabric or paper to break up and remove small areas of the coloured film. This is known as the subtractive method because you are taking away colour to produce the effect
- **Additive method**: Scumble that has been applied to crumpled lint-free fabric or natural sponge is pressed onto the painted ground coat. This is known as the additive method because you are adding colour to produce the effect.

These techniques are covered in more detail on pages 269–270.

Suitability of the ground coat

Before starting to produce any broken colour effect, check the suitability of the ground coat. You need a quality finish that has no visible defects (such as misses, ropiness, bits and nibs, brush marks or excessive orange peel), but you also need to check that the colour is appropriate, that it has good opacity and that the paint film is **hard dry**.

The relationship between the ground coat colour and the scumble colour will influence how pleasing the final result is, and totally different effects will be produced when a pale scumble is applied over a darker ground coat (e.g. light blue-green scumble over dark blue-green ground), as opposed to a darker scumble applied over a paler ground coat (e.g. red-purple scumble over a lilac ground). Whichever order you choose, the two colours should be of a similar tone and not too strongly contrasting.

Materials

It is important to understand the materials used to create broken colour effects, which are oil based or acrylic (water-borne). Although this chapter discusses how to use different products and materials, you should always read the manufacturer's information to find out exactly how each one should be prepared and used, as there can be differences between products.

Oil glaze

This is a translucent **medium** containing **linseed oil**, white spirit, extenders, beeswax and driers, which has little or no opacity.

Chapter 4 explains how coatings flow out and dry as a smooth, level film and that this is directly related to their viscosity. However, if you are producing a broken colour finish, you do not want the scumble to flow out, otherwise the effect will disappear before it dries. The inclusion of beeswax in oil-based glaze enables the scumble to retain its shape (or pattern) after it has been manipulated.

Because oil glaze is based on linseed oil, it has a natural tendency to yellow. This happens over time, but the effect is accelerated on surfaces that get hot, such as radiators, or where there is little light, for example behind pictures, furniture and so on. Potential customers should always be advised of the possibility of this happening, so they may make an informed decision on the range of products that are available.

▲ Figure 6.62 Raw linseed oil can extend the working time

Preparation

Oil glaze should always be thinned using white spirit (following the manufacturer's instructions) so that it can be applied as a fairly thin coat. Take care, though, not to over-thin it, as this will affect its ability to hold the pattern. Once it is the right consistency, colourant is added.

Working time

Oil glaze has a long 'open' or working time, which may be an advantage, particularly when you are working on large areas. However, the working time of any coating will be affected by the atmospheric conditions (e.g. heat, cold, damp) and you need to know how to adjust the material accordingly:

- Higher temperatures will shorten the drying time. Adding a small quantity of raw linseed oil will extend the working time.
- Lower temperatures will lengthen the drying time. Carefully adding a very small quantity of Terebine driers will speed up the drying process, which reduces the working time. However, if too much is added, the scumble will become brittle (cracking and losing adhesion) and the life of the decorative effect will be reduced.

When calculating the quantity of scumble you need for the size of area to be worked, always prepare more than is required. This will enable you to:

- produce a sample board for the customer
- rub out and apply again if the effect is not as required or gets damaged while wet
- leave some scumble with the customer on completion of the job, for any minor repairs that may be required, or to enable a small area to be produced to match as closely as possible to the existing finish, for example if a radiator or pipework is replaced.

Bear in mind that it is difficult to repair damage perfectly. If the scumble is oil-based, yellowing may already have started and so an identical match will be impossible.

Storage

Oil glaze is flammable, so it must be stored in cool, well ventilated conditions, with the lid of the container secured. When working indoors, glaze and thinners should ideally be stored in a metal, fireproof cabinet.

ACTIVITY
Find out the type of fire extinguisher to use on a solvent-borne or oil-based coating fire.

▲ Figure 6.63 Store flammable liquids safely

Acrylic glaze

This is a liquid coating made up of small particles of acrylic resin dispersed (scattered) in water. It is milky in appearance, but when the water evaporates the particles coalesce to form a continuous, translucent film with a slight sheen. (For more about the drying process of emulsion paints, see Chapter 4.)

Manufacturers produce a range of acrylic glaze types, and the product information (technical data and safety data sheets) will help you select the most suitable type for the work. These products have a low odour and a working time of up to one hour. Tools are cleaned with water, and these glazes will not yellow over time. However, they may be prone to chips or scratches and are therefore most suitable for application to walls rather than woodwork. Because acrylic scumbles dry quite quickly, a higher skill level may be required to produce a quality broken colour effect on large areas.

It is important to remember that glazes, both oil-based and acrylic types, do not contain any colourant, and they are not intended to be used as a clear protective film.

▲ Figure 6.64 Scumble acrylic glaze

Preparation

The only preparation required for acrylic glaze is to add an appropriate colourant (see below) and adjust the consistency of the material to suit the atmospheric conditions and required working time.

Working time

Because it is a water-borne product, acrylic glaze has a shorter **working time** than an oil-based glaze. The challenge for the decorator is often how to extend the working time, particularly when applying it to larger areas or in warm weather.

Decorators may use a number of methods to extend the working time of acrylic scumbles, which are not necessarily approved of by manufacturers. These include the addition of a small quantity of glycerine, or lightly spraying the surface with water before applying the acrylic scumble. While proprietary conditioners (which some manufacturers can supply) may help to maintain a wet edge for water-borne products, they also improve the flow of the material and help to reduce application marks. This latter characteristic is not a desirable quality for broken colour work, so if you do use a proprietary conditioner, take care to add a suitable quantity.

Using a wet rag rather than a dry one will help to break up the scumble and will also help to extend the working time of acrylic scumble.

One manufacturer advises that, if using emulsion paint as the colourant, the working time may be altered according to the ratio of acrylic glaze to emulsion paint. For example, 8:1 may give one or two hours of working time, while 4:1 will give about 45 minutes.

As time goes on, manufacturers are improving the open time of acrylic glazes and there is less need to use additives to extend the working time.

Storage

This water-borne product should be stored in cool, well-ventilated, frost-free conditions.

Scumble

This is glaze (either oil based or acrylic) to which colourants have been added, making a new material – a translucent coloured glaze.

It is easy to get confused about the correct name for these different products, particularly when some manufacturers call their glaze 'scumble glaze'. Remember the following:
- If the product is called 'glaze' or 'scumble glaze', it does not contain any colourant and is not suitable to be used on its own as a finish.
- If the product is called 'scumble', it contains colourant and is ready to be used for broken colour work.

▲ Figure 6.65 Scumble contains colourant

Consistency

The scumble shouldn't be thick and sticky, but neither should it be too thin, otherwise its ability to hold an attractive decorative finish will be reduced. Broken colour effects should only have 'visual texture' which is two-dimensional and smooth, not 'tactile texture' which is three-dimensional and can be both seen and felt.

263

As well as understanding the difference between scumble and glaze, you also need to understand the difference between coatings that are translucent and opaque. The meaning of translucent has been explained in this chapter (see page 260), because it is an important characteristic of the glazes and scumbles already discussed. 'Opaque' means a solid colour finish, which obliterates the surface beneath. Paints and some timber treatments are opaque coatings.

Colourants

Good-quality colourants that are stable, lightfast and compatible with the particular glaze should be used. Because the glaze is translucent, even a small amount of colourant will appear as a strong colour. However, you need to add sufficient colour to ensure that the glaze is tinted as strongly as possible and will give consistent colour when applied thinly over the ground coat. Even though the glaze may appear very strong in the kettle or pot, when it is brushed out across a larger area it will be less **intense**.

KEY TERM

Intense: extreme, very strong or having a high degree of colour.

If you are using tube colour, it is helpful to place a strip of the colour on a palette and then mix in a small quantity of the thinner (white spirit or water). This softened colour will be easier to mix into the glaze.

▲ Figure 6.66 Acrylic paints can be mixed with a glaze to create a scumble

Whatever type of colourant you use, take care to fully mix and evenly disperse it throughout the glaze;

if necessary, strain the scumble (using a fine mesh stocking or proprietary strainer) to ensure that there are no lumps of pure colour remaining.

Before starting to work, test the depth of colour on a small board or piece of card coated in the ground colour. Remember, you can always add a little more colour, but can never take it away.

The following colourants may be used in oil-based glaze:
- artists' oil colours
- solvent-borne **eggshell** paint
- universal stainers.

The following colourants may be used in acrylic glaze:
- artists' acrylic colours
- proprietary acrylic colourants
- emulsion paint
- universal stainers.

For training purposes, poster colours may be used in acrylic glazes. Also note that some manufacturers advise against using acrylic eggshell as a colourant.

If solvent-borne eggshell or emulsion paint is used to colour a glaze, remember that the pigments used in those coatings are designed to obliterate the surface, a characteristic that is not desirable in the scumble – you should therefore use them very sparingly.

KEY TERM

Eggshell: a type of semi-gloss paint coating.

ACTIVITY

Name a colourant type that may be used in both oil-based and acrylic glazes.

Tools and equipment

Refer to pages 256–257 for information about paint brushes, hair stipple brushes, mohair rollers, kettles and plastic pots.

Lint-free cloth

'Lint-free' means a cloth without fluff or loose fibres, for example cotton sheeting or mutton cloth. The texture of the cloth you choose will determine the effect produced.

Other products that may be used to produce broken colour effects include paper (tissue/kitchen paper), net curtain material, plastic film or bags. Any cloth used must be lint-free, or the loose fibres will stick to the wet scumble, which is unsightly.

▲ Figure 6.67 Paper towel can be used to produce a broken colour effect

Precautions

Cloths that have been used with oil-based scumbles pose a fire risk – they may **ignite** (catch fire) spontaneously. When not in use, these cloths should be:

- laid flat on a surface to dry by allowing the solvents to evaporate, or
- placed in a metal bin with a cover, or
- immersed in a bucket of water.

You should always take these precautions during lunch breaks or other short breaks, as well as at the end of the day.

Chamois leather

This type of leather is soft, supple, absorbent and non-abrasive. Although imitation chamois (pronounced 'sham-ee') leathers or synthetic chamois leathers are available, they do not have quite the same natural qualities as the genuine article, which is more expensive.

If the chamois leather has been used for oil-based scumble, clean it in white spirit first (it is essential to wear gloves for this) to remove all traces of scumble. If an acrylic scumble has been used, rinse the leather thoroughly in warm water. Using a mild bar soap (not detergent, which will cause the chamois to become dry, brittle and less absorbent), lather the chamois, rinse it out and lather again with the bar soap, but do not

rinse. With the soap still in the chamois, squeeze it dry and gently stretch it out; hang the chamois in an area protected from direct heat and sunlight.

▲ Figure 6.68 Chamois leather

Just before you use the chamois next time, rinse it in warm water to remove the remaining soap, and squeeze out the water. Leaving the soap in the chamois between uses will help to keep it conditioned and preserve its qualities.

Chamois leathers must not be placed in a sealed plastic bag or other container while still wet, as this will cause them to break down and make them unusable.

Natural sea sponges

Sea sponges come from animals that live on the ocean floor; of the 5000 sponge species, only seven are harvested and used for decorative, craft and cosmetic purposes. Sea sponges are soft and pliable, and their texture enables the creation of random and unique patterns. They may be used with oil-based, water-borne and acrylic coatings.

▲ Figure 6.69 Natural sea sponge

Care of sea sponges

As with cleaning a chamois leather, the first stage of cleaning a natural sea sponge will depend on the type of scumble used – this should be followed by thorough rinsing in warm water until it runs clear. A sea sponge must be handled carefully. Gently squeeze it rather than wringing it out as you would a cloth, leather or synthetic sponge. Sea sponges are expensive to buy and should be dried before being stored in a location where they will not be damaged.

Dragging brush

This brush is used to produce specialist finishes and is usually associated with **graining**, but it may also be used to produce a broken colour effect (see pages 271–272 earlier in this chapter). The filling has either two rows of natural bristle, or one row of bristle and one row of stiff nylon, which replicates the split whale bone used in the past. The filling type influences the effect produced, with those containing nylon producing a coarser effect than the bristle-only type. This tool is not used to produce any of the broken colour effects required in this chapter.

▲ Figure 6.70 Dragging brush or graining drag

KEY TERM

Graining: applying and manipulating an appropriately coloured scumble to imitate the appearance of a specific timber.

Care of the dragging brush

The dragging brush is cleaned using the appropriate thinner for the material it has been used with. When all traces of the thinner have been removed, wash the brush in warm, soapy water and rinse it. It should be allowed to dry fully before being stored, preferably in a box to protect it against damage.

Palette

▲ Figure 6.71 Paint palette

This is a timber or plastic board (sometimes with a hole to place the thumb through) on which a quantity of paint is placed.

Setting out

You may be working from a scale drawing (perhaps showing panel areas on a wall), from verbal instructions, or you may be applying an effect to a specified architectural area such as a dado. If required, measure and mark out the area that is going to receive the broken colour effect, using appropriate tools such as a tape measure, a **chinagraph (or grease) pencil**, a spirit level for horizontal lines and plumb bob for vertical lines, and chalk and line (or self-chalking line). Check the dimensions for accuracy throughout the setting-out process, and then apply masking materials.

KEY TERM

Chinagraph/grease pencil: a writing implement made of hardened coloured wax that is useful for marking on hard, glossy non-porous surfaces.

Protecting adjacent areas: masking tape

▲ Figure 6.72 Masking tape

When you are producing decorative effects on walls that are next to the ceiling or adjoining walls, it is necessary to protect the surrounding surfaces from damage caused by the materials and tools used.

For the protection to be effective and not cause damage, you will need to select the appropriate kind of masking tape by considering:

- the type of surface to be protected (smooth, delicate, rough, etc.)
- the length of time it will need to be protected (one day, a week, etc.)
- the material you are protecting against (solvent, water, etc.)
- whether it is for an interior or exterior surface (consider temperature and weather, including **UV light**).

KEY TERM

UV light: ultraviolet rays, usually from the sun, which can cause health problems (e.g. with the skin or eyes) and damage to materials.

Masking tape should:
- produce perfect, sharp edges
- hold straight lines over long stretches
- make precise curves
- be removed cleanly, without leaving residue or damaging the substrate.

Adhesive tape is made up of the backing (which may be paper, fabric or plastic film) and an adhesive (rubber based, acrylic based or silicone based). Different combinations of these materials produce a range of tapes. Tape is rated by how many days it may be left on a surface without leaving a residue when removed. The longer-rated tapes have less adhesive and are most suitable for smooth, delicate surfaces, such as a recently painted ground coat.

ACTIVITY

Find out which type of adhesive tape is most suitable when masking curves.

Application

If you are using separate masking tape and paper, the tape will need to be adhered to the edge of the paper, in manageable lengths. Alternatively, a self-adhesive paper or pre-taped polythene sheet may be used, depending on the location and surface area of the work. First, accurately position the masking material lightly at the edge of the area to be decorated, then apply even pressure to the tape so that it firmly adheres to the surface. This will help to prevent any **creep** of scumble under the edge of the tape, which will lead to thick, blurred edges.

INDUSTRY TIP

Press the front of a thumbnail along the edge of the masking tape, to ensure that it is really well secured and to prevent creep.

KEY TERM

Creep: where masking tape has not been securely fixed to a surface and some scumble seeps beneath it – this will result in there not being a sharp edge to the broken colour effect.

Removal

Masking materials should be removed as soon as is feasible after the work has been completed. When de-masking, take care to avoid:
- damaging the finished effect, as it is almost impossible to invisibly repair and usually results in the scumble having to be rubbed out and the effect created again

- lifting the ground coat. Should this occur, even shallow damaged areas will need to be filled (using fine surface filler) and spot-primed, before re-coating with the ground coat and feathering in the edges.

When removing masking tape, both the surface and the adhesive tape should be dry. It should not be pulled away from the surface, but instead needs to be pulled back on itself with a careful, even pulling motion. If you do not do this, the backing material may tear and leave adhesive residue on the surface, or the ground coat could be lifted.

▲ Figure 6.73 Take care when removing masking tape

Masking film

Masking film is available in sheets or rolls, for example Frisket Film. This is a self-adhesive film that sticks to the surface and leaves neither residue nor surface damage when peeled from the surface after use. It is easy to cut and has a translucent backing. Some types of masking film are repositionable and, if correctly adhered, will not buckle or lift along the film edge, which would allow creep. A range of tack levels (such as low tack or extra tack) and finishes (gloss or matt) are available.

Masking film may also be applied as a liquid using a spray or roller, for example the Protectapeel or Peelgard brands. The liquid dries to form a tough, skin-tight, plastic film, which will protect a range of surface types. It is water-based and can be sprayed over a design, which can then be cut out, providing masking for the surrounding area. It is easily peeled off and disposed of by recycling or as domestic waste.

These products are more expensive than the masking tape and materials mentioned above and used for specialist applications, such as airbrush work, stencilling or precise colour control.

It should also be noted that masking film must be cut carefully, using minimum pressure, to prevent damage to the underlying surface from the sharp knife that is required to cut through it.

ACTIVITY
Find out what 'frisket' means and where the word comes from.

Removal

Masking materials should be removed as soon as is practicable after the work has been completed. When de-masking, as with masking tape, take care to avoid damaging the finished effect and lifting the ground coat. Should this occur, even shallow damaged areas will need to be filled (using fine surface filler) and spot-primed, before re-coating with the ground coat and feathering in the edges.

Disposal of waste

Removed masking materials are a slip hazard if they are not placed immediately in a bin liner. If oil-based scumble has been used, both masking materials and cloths are a potential fire risk.

Storage

Tapes should be stored in a dry location, at room temperature, protected from direct sunlight and freezing temperatures. In these conditions they can be stored for at least 12 months. High temperatures will speed up the ageing process and will affect the usability of the tape.

Application techniques

Having prepared your scumble (glaze plus colourant) and obtained the required consistency, you are now ready to produce a variety of broken colour effects.

Step-by-step rag rolling: subtractive method

This may also be referred to as the **negative technique** or **ragging off.**

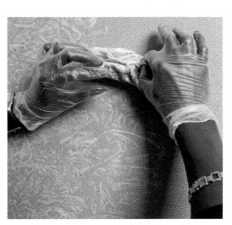

STEP 1 Apply the scumble using a well-worn brush (natural bristle may be used for oil-based scumble, but synthetic filament must be used for acrylic scumble), and rub out to cover the surface evenly, being careful not to apply too much glaze.

STEP 2 Remove all brush marks from the scumble and produce an evenly textured finish using a hair stippler or mohair roller. A stippler is used by firmly striking the glaze at 90° with the bristle tips, using short, sharp, clean strokes, and moving it slowly across the surface until an even texture is achieved.

STEP 3 Crumple a piece of lint-free cloth (or chamois leather) and firmly but lightly roll this across the area, working in random, snake-like moves, slightly overlapping each time to avoid unrolled areas or tramlines (this is also called banding or tracking).

INDUSTRY TIP

If you are using an acrylic scumble, dip the cloth in water and wring it out before starting. This will help to break up the scumble and extend the working time.

Before using a stippler or roller for the first time, prime the bristle tips or roller sleeve by drawing the rubbing-in brush (see page 280) across them a few times. This will help to reduce the porosity of the bristles' oblique pile and keep scumble from being absorbed by them, which could lead to areas of thin or removed scumble on the surface you are decorating.

KEY TERM

Discernible: clearly seen.

When you perform Step 2, if the pressure of the stroking action is too light, it will take a long time to remove the marks and even out the scumble. However, if too much pressure is applied there can be a tendency to drag the tool across the surface, and this will leave slip or skid marks in the scumble.

You should repeat Step 3 until the entire surface has been rolled and an even effect achieved, with no **discernible** pattern. If a more detailed effect is required, either use a thinner type of cloth that will crumple with more broken edges or roll the entire surface again to lift off more glaze.

INDUSTRY TIP

If working an area or section of 1 m² or more, it is advisable to apply the scumble with a roller, which will combine Steps 1 and 2 and be more time and cost effective.

At regular intervals, shake out and re-crumple the cloth, taking care not to fold it, because the more broken edges there are in the ball of cloth, the more interesting the pattern. To prevent the cloth becoming overloaded with scumble, periodically rinse it out in the appropriate thinner (white spirit or water), then wring it out well.

When working on large areas such as walls, consider the following:

- It is advisable for one person to apply and stipple the scumble, and a second person to create the effect. As each person's work is unique to them, the film thickness and pressure used when applying and stippling the scumble, and the way of crumpling the cloth and method of manipulating it, will also be unique. Roles should not be reversed partway through the job, as the final effect will be quite different and noticeable.
- It may be necessary to replace the cloth being used. The new one must always be from the same piece of material, as the texture and pattern being produced will be unique to that product.
- Getting into corners or small areas may require using a smaller piece of cloth, yet still producing the same effect.

INDUSTRY TIP

To check that both the colour and effect are even, partially close your eyes and squint at the work – this will clearly show up any uneven patches.

Rag rolling: additive method

This may also be called the **positive technique** or **ragging on**.

1 Place the prepared scumble in a container and immerse the cloth or chamois leather in it then wring it out. It is advisable to press the cloth onto on a spare board or length of lining paper, to avoid applying too much scumble to the surface.
2 Roll the cloth over the surface in an irregular manner, making sure not to double-roll any areas, but also not to leave any part of the surface unrolled, otherwise tramlines or banding will result. If the amount of scumble being deposited on the

surface begins to reduce, re-immerse the cloth in the scumble to re-load it, and wring it out.

▲ Figure 6.74 The additive method of rag rolling

An advantage of the additive method is that there is no wet edge to keep open. However, it can be more difficult to achieve a **uniform** effect than with the subtractive method of rag rolling.

KEY TERM

Uniform: regular; always the same in shape/form, quality, quantity, etc.

Sponge stippling

This is also an additive method or positive technique. This decorative effect gives a multi-coloured, speckled appearance, with a minimum of two colours being applied to the ground coat, and three colours giving an extra dimension to the work.

An advantage of using emulsion paint for this effect is its speed of drying, particularly as two or three colours are being applied on top of one another and each colour must be dry before the next one can be applied. A large natural sponge should be used, particularly when working broad areas such as walls. Prepare it for use by immersing it in water and gently squeezing it out.

1 Place emulsion paint or prepared scumble in a roller tray or large dish.
2 Press one side of the sponge into the paint or scumble to load it.
3 Dab the loaded sponge on a spare board or length of lining paper to avoid applying too much to the surface. Alternatively, use thinned emulsion in a container, immerse the sponge and squeeze it out, then dab off the excess paint.

4 Stipple the paint onto the wall using a light dabbing action, turning the wrist (not elbow) each time to avoid any regular pattern being created. Take care not to drag or twist the sponge on the surface, otherwise skid marks may be created. The stipples should be placed close together, but not overlap.

5 When the first colour has dried, apply the next colour(s) in the same manner, always applying the lightest colour last of all. If minor damage should occur, apply a repair using the same colour as the ground coat to disguise it.

▲ Figure 6.75 First colour applied overall

▲ Figure 6.76 Second colour applied

IMPROVE YOUR MATHS

A room with dimensions of 5.26 m × 3.5 m × 2.45 m high, with windows and door area of 4 m² is to be decorated with a rag-rolled effect (see pages 269–270).

Calculate the following:

1 The quantity of scumble glaze required, if the spreading rate is 8 m²/litre.

2 The cost of the scumble glaze @ £33.32/2.5 litres, plus VAT at 20%.

3 The cost of producing a rag-rolled effect on the wall areas, if the unit cost is £6.55/m².

(Answer provided on page 301.)

Dragging

Dragging is a negative or subtractive technique. The decorative effect produced has an irregular striped appearance which is formed by the brush filling producing many thin lines as it is drawn (dragged) across the surface. It may be applied to small areas, for example to outline or frame panels of other decorative effects or cupboard doors, or to broad wall areas. If it is applied to broad areas, these should ideally be without obstructions or architectural features, which cause the effect to be interrupted and therefore make it more difficult to produce. A broken or interrupted appearance may also result when working on high walls, as you will have to descend steps while trying to maintain an even pressure and parallel vertical lines.

As with all decorative finishes, the surface preparation and finish must be of a high quality, otherwise the scumble will form dark blotches in surface indentations.

If the effect is to be applied to large areas, it is advisable for two people to undertake the work, as described for rag rolling (see page 269).

▲ Figure 6.77 A dragged effect produced on a panelled door

Stage 1

Apply the scumble using either a well-worn brush (natural bristle may be used for oil-based scumble, but synthetic filament must be used for acrylic scumble) or, if a broad area is to be worked, apply the scumble using a medium-pile roller. Rub out to cover the surface evenly, ensuring that not too much glaze is applied, particularly at the edges, where a heavy build-up/ deposit can occur. Take care when applying scumble by roller, as the ingredients make the material very likely to produce skid marks from the roller sleeve.

Scumble should only be applied to an area of a size that can be manipulated without losing the wet edge. If working alone on a large wall area, apply the scumble in vertical bands of approximately 1 m and apply the effect to within 150 mm of the leading edge – apply the next band and continue the process, working quickly to keep the wet edge moving.

Stage 2

Before using the dragging brush for the first time, prime the bristles or filaments lightly with scumble, as for rag rolling. The exact brush type used will depend on the size of the area and whether there are any obstructions or architectural features.

Stage 3

Holding the dragging brush at a shallow angle to the surface, pull the brush firmly across the surface in the direction required. Do this without hesitation and in a single, straight stroke if possible, to produce continuous, yet irregular stripes. Repeat this process, slightly overlapping the brush strokes and keeping them parallel.

If you are producing the effect on a broad area such as a wall or a flush door, start at the top and work to the bottom. However, if you are applying it to a panelled door/area, observe the sequence of application used to reflect the component joints – muntins, rails, stiles (see Figure 6.78).

At regular intervals, wipe the bristles with a lint-free cloth to remove the excess scumble which is being taken off the surface – this is necessary to produce a clean, even finish.

▲ Figure 6.79 A dragged effect being produced on a flush panel

Stage 4

When working on an area with adjacent surfaces such as ceiling, skirting boards, floor, frames, etc., care and practice are required to produce the dragged effect, not only right to the edge, but also maintaining the weight of colour (quantity of scumble) and an even effect.

At the bottom edge, reverse the brush and drag it up the wall to blend the new stripes into the existing ones, gradually lifting off the brush. Continue this process along the bottom edge, ensuring that each stroke is a different length to avoid a horizontal band being created.

1. Top rail
2. Stile
3. Muntin
4. Panel mouldings
5. Panels
6. Mid rail
7. Lock rail
8. Bottom rail

▲ Figure 6.78 Panelled door

▲ Figure 6.80 Drag the brush up to correct the fault

On a sheet of card, paper or paper-backed vinyl that has been ground out, prepare and apply scumble and produce a dragged effect, with butt and mitre joints as you would find on a panelled door. Ensure that the 'joints' are crisp and accurate, using a piece of masking tape or coarse abrasive if necessary to protect the scumble on adjacent areas.

At the top edge, apply a small quantity of glaze just beneath the edge and carefully rub out. Angle the brush to get the bristle tips right to the internal angle and drag down to blend into the existing dragged effect, gradually lifting off the brush (as if you were laying off paint).

You will need to follow this process of achieving continuity of effect near skirtings to successfully achieve a uniform dragged effect.

If you are applying the effect to a small area, bear in mind that you can produce a completely different effect by pushing (rather than pulling) a dragging brush through the scumble.

'Glazing and wiping'

This is a negative or subtractive technique that is used to highlight and enhance the following:

- wall coverings that have a textured surface/relief design
- architectural room features, such as ornate covings/cornices or a centrepiece
- surfaces produced by the application of materials that can be textured.

As with all broken colour effects, the relationship between the ground coat and scumble colours is crucial to produce an aesthetically pleasing and effective result – the use of a metallic ground coat paint can be quite effective for this process.

If the 'glazing and wiping' is being applied to large areas, it is advisable for two people to undertake the work, as for rag rolling.

Stage 1

Apply the scumble using either a well-worn brush (natural bristle may be used for oil-based scumble, with synthetic filament being preferred for acrylic scumble) or, if a broad area is to be worked, a medium-pile roller. The scumble must be worked into all the recessed areas of pattern or moulding, followed by the removal of any application marks (e.g. careful laying off by roller, or brush marks by the hair stippler/roller).

As with Stage 1 of dragging (page 271), scumble should only be applied to a size of area that can be manipulated without losing the wet edge. If you are working alone on a large wall area, apply the scumble to sections of approximately 1 m² and apply the effect to within 150 mm of the leading edge, then apply to the next section and continue the process, working from the edge of the area just completed.

Due to the texture of the surface being scumbled, it is best to apply the scumble (of either type) using a medium-pile roller. This will not only ensure that the scumble is applied as a full coat, but will also have benefits when applying acrylic scumble, which has a quicker setting time.

Stage 2

Using lint-free cloth, fold the material to form a pad with a smooth surface (no creases). For large areas such as ceilings, the wiping can be carried out using a stiff rubber squeegee, wiping off the excess scumble with a rag after each stroke; this can be returned to the pot and reused, if uncontaminated.

▲ Figure 6.81 'Glazed and wiped' effect on a relief texture wall covering

Stage 3

Methodically wipe over the surface to remove the scumble from the raised sections of the pattern. Do not press too hard, otherwise the scumble may also be removed from recesses, but the same amount of pressure needs to be applied over the entire area to achieve a uniform effect.

Regularly re-fold the cloth to avoid it becoming saturated, in which case it will be unable to remove the scumble. It may be necessary to replace the cloth being used – the new one must always be made of the same type of material, to ensure the same effect is achieved.

▲ Figure 6.82 'Glazed and wiped' effect on a frieze

If insufficient glaze is removed, repeat the process, but at 90° to the direction of wiping used the first time, or increase the pressure you are using to wipe the surface. It is advisable to work on a test area before starting on the surface, to familiarise yourself with the material. Bear in mind that atmospheric conditions can have a great effect on surface coatings, and a material used on one day may handle differently the next if the conditions have changed.

The use of oil-based scumbles means that more lint-free rags will be required – these can be rinsed out in white spirit if required (wear gloves to protect your skin when using white spirit).

If you are using an acrylic scumble and a longer working time is required, dip the cloth pad in water and wring it out well, which will help to remove the scumble. Rags can be rinsed out and reused, which will mean fewer contaminated rags to be disposed of.

HEALTH AND SAFETY

This decorative effect requires you to hold the lint-free rags in your hands. Wear gloves to protect your skin from harmful ingredients in both water-borne and solvent-borne scumble types.

Application faults

Banding/tracking

Banding or tracking can occur if you don't slightly overlap areas while rag rolling or check the work carefully on completion.

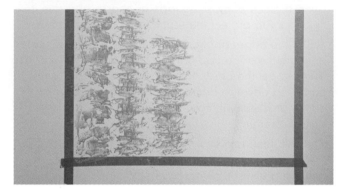

▲ Figure 6.83 Ensure you overlap areas when rag rolling to avoid tracking

Slip/skid

This fault can occur if you apply too much pressure or don't take enough care when evening out the scumble or producing the effect.

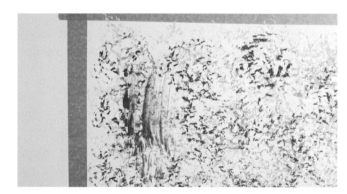

▲ Figure 6.84 Slip/skid in a broken colour effect

Creep

Creep occurs when masking tape has not been securely fixed to a surface and some scumble seeps beneath it. This will result in there not being a sharp edge to the broken colour effect.

Surface damage

Masking materials should be removed as soon as is practicable after the work has been completed. When de-masking, take care to avoid damaging the finished effect and lifting the ground coat.

Protection of the finished work

Broken colour finishes are delicate, and when they have been applied to areas that are subject to knocks and damage, such as doors, woodwork and certain wall areas, they need to be protected with a varnish that is compatible with the type of scumble used. An eggshell or flat varnish finish is most appropriate, as the light reflectance from any higher degree of sheen will detract from the decorative finish. The varnish should be applied as thin coats and built up, according to the durability required.

ACTIVITY

Using the internet and manufacturers' information, find a non-yellowing varnish suitable for protecting oil-based and acrylic scumble work.

⒊ PRODUCE FAUX DECORATIVE EFFECTS

Producing wood and marble effects using basic techniques

The art of graining and marbling, which is the imitation of woods and marbles using paints and scumbles, can be traced back centuries. It became popular in Britain in the mid- to late 1800s, as a substitute for the expensive timbers being brought into the country with the expansion of the British Empire. The rich were able to access foreign marbles to create homes that displayed their wealth, and everyone else wanted to imitate these.

INDUSTRY TIP

Many people are unclear about the difference between wood and timber. In simple terms, wood refers to the tree itself and timber refers to the wood at any stage after the tree has been cut down.

To create a realistic imitation of woods and marbles, it is necessary to:
- understand their origins – the methods used to process them following felling and quarrying, and the factors that influence the pattern structures and colours of each
- consider the location they will be used in, as they will be aesthetically pleasing only if they are used realistically and in context. For example, it would be inappropriate to produce a grained door with marbled mouldings, or grained iron railings or rainwater pipes.

As with all broken colour effects and decorative work, the requirement for a high-quality ground coat of an appropriate colour is essential and involves thorough surface preparation.

INDUSTRY TIP

This section looks at the materials, tools and processes used to produce basic graining and marbling effects. If you are interested in this specialist area, the following are useful sources of information.

Parry's Graining and Marbling by Brian Rhodes (1995) Oxford: Wiley-Blackwell.

The Art of Marbling by Stuart Spencer (2000) London: Little, Brown.

The Paint Effects Bible: 100 Recipes for Faux Finishes by Kerry Skinner (2012) Richmond Hill, ON, Canada: Firefly Books.

www.youtube.com – search for 'Introduction to Marbling and Graining' by Stewart McDonald (in conjunction with Dulux Trade UK).

Ground coat colours

For graining, the choice of ground coat colour often comes down to personal preference. However, the decision will be based on an understanding of colour technology, the characteristics of the wood or marble you want to imitate and the various pigments/colourants which can be used in the process. Experienced grainers often mix their own colours to obtain the required tones, particularly if they are matching existing work. Under these circumstances, the lightest colour of the existing work will be matched for the ground coat but may still require a little adjustment when a sample board is produced.

The **integrity** of a wood or marble imitation may be compromised by the use of an inappropriate colour for the ground coat. However, when producing fantasy (imaginary) marbles, achieving an imitation of real-life marble may be less important than the ability to produce an effect that fits the desired colour scheme.

> **KEY TERM**
>
> **Integrity:** truthfulness or accurate characteristics.

The British Standard 4800 colours for wood and marble effects are detailed below:

- BS 4800 colours for graining:
  - Light Oak – tint of 08 C 35
  - Medium Oak – 08 C 35 (or 06 C 35). The description 'buff' coloured or 'biscuit' coloured is sometimes used
  - Mahogany – 04 D 44 (or 06 C 33).
- BS 4800 colours for marbling:
  - Carrara – 00 E 55
  - Fantasy (a black and green marble type is the example used) – 00 E 53.

▲ Figure 6.85 Medium Oak

Oil-based products used in graining and marbling materials

Oil glaze

When graining with oil colour, a small amount is used with the pigment, to ensure that the initial rubbing in holds up (retains its shape) while being manipulated, for example by being combed or having **figurework/figure graining** applied.

Oil graining colour

This medium (previously known as **megilp**) is applied at the initial rubbing-in stage (see page 281). Traditionally it was made from one part linseed oil, two parts turpentine (derived from the resin of pine trees), liquid driers and pigment.

Nowadays, oil glaze rather than megilp is used, and appropriate pigments (which are ground into oil at the manufacturing stage) are added to obtain the required colour, with white spirit added to obtain the required consistency (viscosity).

> **KEY TERMS**
>
> **Figurework/figure graining:** the main design or pattern of a wood grain, which differs between tree species and according to the **method of conversion**.
>
> **Megilp:** a variety of substances such as soft soap or beeswax that used to be added to graining colour to make it hold up or prevent it flowing together after it had been combed or figured.
>
> **Method of conversion:** the way a newly felled tree is sawn up into usable-sized pieces of timber, using two main methods: through-and-through or quarter-sawn.

> **ACTIVITY**
>
> Using your college or training centre's resources, find out more about the two methods of converting timber, and produce a drawing of each method.

Oil-based proprietary scumbles

These are ready-made oil graining colours, coloured appropriately to match the most common range of woods. They must be stirred and thinned prior to application and, although ready-made, the colour can be adjusted by the addition of oil colourants to meet individual requirements.

> **INDUSTRY TIP**
>
> The drying time for oil-based scumbles can be extended by the addition of small quantities of linseed oil. However, take care not to add too much, as this will mean the scumble will not dry properly.

Varnish

This is a solvent- or water-borne clear coating that contains film former and thinner, but no coloured pigment. It is necessary to apply varnish to protect graining and marbling effects because the materials used to produce them have little protective quality and are not very resistant to abrasion. Remember that the varnish needs to be compatible with the materials used and should have a level of sheen appropriate for the work. For example, polished marble requires a full-gloss varnish and waxed wood requires an eggshell sheen.

> ### INDUSTRY TIPS
>
> Eggshell and flat varnishes contain a flatting agent which reduces light reflection, and so they should be stirred before use. Gloss varnish should not be stirred, but gently shaken just before use.
>
> Rags soaked with linseed oil and left in a pile are a fire hazard. The linseed oil oxidises leading to an exothermic reaction (a reaction that releases energy in the form of heat) and this can lead to spontaneous combustion.

Driers

For more information about driers, see page 277.

Colourants and pigments used in graining and marbling materials

For further information on colourants see page 264.

Crayons (oil pastel/wax oil crayon)

These are sticks of pigment mixed with a non-drying oil and wax binder that are used for graining – to put in figurework or figure graining, and marbling – for fine veins.

They may be used:
- in a dry form, and once applied to the surface they can be manipulated with a brush moistened with white spirit or linseed oil
- by oiling in the surface prior to applying the detail with crayon
- with oil colour work and water colour, but should be tested before varnishing, as they may be difficult to protect.

Gouache

This is a type of paint that is similar to watercolour but modified to be opaque. It consists of pigment and binder, sometimes with an extender such as chalk added. Gouache has good reflective qualities but generally dries to a different tone than it appears when wet (lighter tones generally dry darker, while darker tones tend to dry lighter). It is available in many colours and is usually mixed with water to achieve the desired working consistency and to control the opacity when dry.

Universal stainers

These are high-strength, concentrated stainers, which may be mixed with most types of paint.

> ### INDUSTRY TIP
>
> When you are working with acrylic scumbles, keep a small pot of methylated spirits handy to soften any brushes used. Badger-haired softeners, in particular, are easily contaminated and expensive to replace. Dry brushes should be immersed in the spirits, which will soften the acrylic, allowing them to be washed in soapy water and rinsed.

Water-based materials used in graining and marbling materials

Water-based and acrylic colours

Water-based graining colours are reversible when dry, which means they can be softened, or re-activated with water, or by applying more of the graining colour. For more information about acrylic glaze and acrylic colourants, see pages 262–264.

Dry (powdered) pigments

The most common pigments used are 'earth' colours and include **ochres** (pale yellow to deep orange/brown), **siennas** (yellow/brown) and **umbers** (brown or reddish brown). These are naturally occurring minerals, mainly iron oxides, which are found in rocks and soils. Some earth pigments are roasted in order to intensify their colour and are then known as 'burnt', as opposed to 'raw'.

Because of their good opacity, only small quantities of earth pigments should be used, to ensure that the scumble retains its translucency.

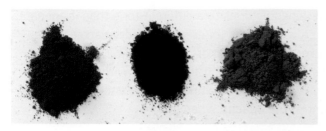

▲ Figure 6.86 A range of powdered earth pigments

Binders or fixatives (for water graining colour)

These are used to temporarily bind or hold together dry (powdered) pigments while they are manipulated to form the grain pattern required. A varnish coat will be needed to permanently bind the dried water-borne scumble to the surface.

Fuller's earth

Fuller's earth is a naturally occurring sedimentary clay, used in a fine powder form. It is mixed with water and is used as:

● a binder/fixative for pigments
● to prevent cissing – a damp sponge or cloth is dipped in the powder and rubbed over the surface, which is then rinsed off.

▲ Figure 6.87 Fuller's earth

Whiting

This is powdered and washed white chalk (calcium carbonate) which is sometimes added to paint as an extender to provide body. It may also be used to prevent cissing, but not as successfully as Fuller's earth, though it is applied in the same manner.

Stale beer and malt vinegar

These were traditional binders used by grainers, as they were readily available and economical and are commonly still used today.

Glue size

This is a binder for pigments made from animal glue. The pigment colours appear matt and opaque, with reds and blues appearing quite strong, as opposed to having a translucent appearance when used in oil.

Retarding agents
Glycerine

The addition of glycerine to water graining colour slows down the evaporation rate of the water, allowing an extended working time. However, if too much is added, it will stop the drying process, so only a very small quantity should be used.

Materials used to prevent cissing

If you apply water-based materials to an oil-based or solvent-borne coating, you will probably experience cissing (see page 282). The following materials work by reducing the **surface tension** (a form of degreasing), allowing the water-based material to adhere as a continuous film. Detergent (a **surfactant**, such as washing-up liquid), Fuller's earth and whiting can all be used to prevent cissing.

Graining and marbling tools

Specialist tools are available that are specifically designed for the highly skilled techniques used in graining and marbling. These are expensive, so particular attention should be given to their care and maintenance. However, it is possible to achieve a good standard of work with everyday decorating tools, and even some **improvised** items.

Combs

These may be made from metal (steel), rubber or card (improvised) and are used mainly for oak graining. Metal combs may be fine, medium or coarse and come in four widths. Rubber combs are either triangular with fine, medium or coarse teeth, or double-ended (3-inch and 4.5-inch sides) with graduated teeth.

▲ Figure 6.88 Metal and rubber graining combs

Check/tick roller

This is a small roller made of serrated zinc discs which revolve independently and take scumble off a **mottler** (positioned above the discs and clipped in place) and then deposit a random series of dark tick marks to simulate the dark open pores of oak grain.

▲ Figure 6.89 Check/tick roller and mottler

Feathers

Suitable types of feather to use include goose, swan and duck, with the goose feather (obtained from the wings) being the most suitable because it has the best shape, size and strength.

Feathers may be used in:
- marbling, to produce background texture/colour, and/or veins
- graining, to produce wavy markings for woods such as walnut.

To care for feathers, regularly wipe off excess material with a lint-free rag during use. When you have completed the work, clean them using the appropriate thinner, then wash in soapy water and rinse – allow to dry fully before storing.

Palette knives

These have carbon steel flexible blades and hardwood handles and come in a range of blade sizes (4-inch, 6-inch, 8-inch, etc.). They are used to mix ingredients, particularly on a palette board.

▲ Figure 6.90 Palette knife

Rags and sponges

See pages 264–265 for more information on lint-free rags and natural sponges.

Rubbing-in brushes

These are part-worn flat paint brushes used to spread (or rub in) the scumble (or medium) over the ground coat. Separate sets, consisting of 25 mm and 50 mm sizes, should be kept for water-borne and oil-based materials.

Mixing brushes

These are part-worn brushes (25 mm is ideal) used to combine ingredients ready for application.

Fitch brushes

These are long-handled, hog hair brushes, available with round or flat filling shapes of varying diameters/widths. They are normally used to produce pattern or figuring detail when undertaking marbling and graining.

▲ Figure 6.91 Fitches – round and flat

Flogger brushes

These are brushes with less filling, but a longer length out (approximately 120 mm), used to produce the pore markings seen in many hardwoods, before the main figuring is done.

Use the flat of the brush (not so much the tip) to strike or 'flog' the scumbled surface, working from the bottom upwards. They are suitable for either oil-based scumbles or water-based colours.

Clean flogger brushes using the appropriate thinner, then wash in soapy water and rinse. Allow to dry fully before storing, to prevent damage.

▲ Figure 6.92 Flogger brushes

INDUSTRY TIP

When used with water-borne scumble (not acrylic), it is advisable not to wash out the flogger brush, as any pigment adhering to the bristles will absorb water from the scumble, making it dry more quickly.

Softeners

These are used to blend or soften sharp edges in both graining and marbling. Use them to create an overall softened effect, by lightly stroking the glaze back and forth in one direction with the tips of the brush filling just touching the surface. When a direction of glaze appears, stroke the surface at 90° to this first direction. Continue until the desired effect is achieved.

There are two types of softener:

- **Hog hair, for use in oil-based medium**: Clean regularly during use (with white spirit) then dry off on lint-free rag before reuse. On completion of work, clean with white spirit, then wash with soapy water and rinse – allow to dry fully before storing, to avoid damage.

- **Badger hair, for use in water-based medium**: Clean infrequently during use, but immediately wash the badger softener well in cool soapy water once finished. Rinse and hang the brush to dry on completion of work.

It is important to note that because acrylic materials dry quickly, the hair needs to be moistened (kept wet) while in use. Should these materials dry in the brush, clean carefully but thoroughly with methylated spirits.

▲ Figure 6.93 Softeners – badger and hog hair

Sign-writing or sable pencils

These are often called 'writing pencils'. They are long-handled brushes, usually with sable or nylon filling held in a metal ferrule or quill (the base of a bird's feather), with a chisel (flat) or pointed end. They are available in a variety of sizes relating to their diameter and length out. They are used by sign-writers to produce hand-painted signs.

Sable pencils are used in graining to produce grain markings and enhance grain characteristics, and in marbling for the application of veins or 'fractures/cracks'.

▲ Figure 6.94 Sign-writing pencils/writers

Carefully clean sable pencils in the appropriate thinner (white spirit or water), then apply a small quantity of petroleum jelly or grease to hold the filling together and straight. They are usually stored in an oblong box/tin or tube. Lightly swill out in white spirit to remove the grease before use.

Varnish

Varnish brushes, whether natural bristle or synthetic filament, should never be used to apply paint, but should be kept as a separate set for varnishing purposes only. It may be beneficial to work brushes into a small amount of varnish prior to application, to avoid any frothing or milkiness in the finished appearance.

Clean out varnish brushes using the appropriate thinner (white spirit or water). Wash them in detergent, rinse in clean water and allow them to dry. Alternatively, for brushes used with solvent-borne varnish, suspend them in a mixture of linseed oil and white spirit, which will help to keep the bristles supple. When required for use, rinse out in white spirit and spin well before use.

INDUSTRY TIP

When using water-based scumbles for graining, it is advisable to bind them down using yacht varnish. The more common polyurethane varnishes are not as resistant to water when **overgraining**.

KEY TERM

Overgraining: the process of simulating subtle lighter and darker areas of the grain as seen in natural wood – this is carried out, usually in watercolour, over completed figurework.

Graining and marbling processes

Rubbing in

'Rubbing in' is the initial spreading of the graining colour, for example using a 25 mm brush for window frames and door mouldings and a 50 mm brush for larger areas. The brush is loaded **sparingly**, with just enough colour to make a thin, even coating on both flat and moulded surfaces. This process is also used in marbling.

KEY TERM

Sparingly: using very little.

281

Oiling in

The term 'oiling in' mainly applies to marbling. It is the process of moistening the ground coat using some of the medium (oil glaze or acrylic glaze) to enable the blending of colour washes.

Flogging

This is where thinly applied scumble or watercolour is struck using a flogger brush, working upwards or away from the body, to produce the tiny pores or fine grain marks seen in most woods. This process is carried out before the main figure, or pattern of the wood, is produced.

Combing

This is mainly used to imitate the hard and soft grain patterns of wood, particularly oak, with different comb types and sizes producing different effects. Combs should always be wiped clean after each stroke, to keep the work clean and prevent a build-up of colour into ridges. In the case of metal combs, when a piece of lint-free cloth is placed over the end of the teeth to soften or blur the effect, this should be frequently moved to present a clean edge.

Veining

This is the process of imitating the veins, fine lines and cracks that occur in marbles, using a variety of tools and materials such as feathers, writing pencils, crayons and so on.

Softening

As its name implies, this process is the softening or fading out of harsh lines that may be produced when manipulating graining and marbling colour. It also enables colours to be blended together. Traditionally this has been achieved using a hog hair softener for oil colours and a badger softener for water colours. The softener is used by holding the brush at 90° to the surface and very lightly brushing the tips of it across the surface; it may be used either with a back-and-forth movement or in one direction only, depending on the desired effect.

Glazing

This is the application of a thin, transparent wash of colour to give a subtle variation in the overall colour when marbling. It is done before varnishing the completed work and must not be carried out until the existing work is hard and dry.

Cissing or opening out

This is achieved by flicking tiny droplets of an appropriate thinner (white spirit or water) onto wet glaze, which breaks up the glaze to reveal rings of the ground coat. This may be achieved by dipping a small fitch brush into the thinner and drawing a forefinger over the bristle tips to produce a light spray. If too much thinner is applied, lightly dab the surface with a lint-free rag to prevent it spreading further.

Wiping out

This is the process of lifting or removing areas of wet scumble to reveal the ground coat or underlying effect and produce figure graining and veins. This is opposed to the painting in or application of these features.

Stippling

For more about stippling, see Step 2 on page 269.

Graining and marbling effects

While you are learning how to produce graining and marbling effects, you should be brushing out your work and repeating it a number of times, to gain practice and experiment with different materials and tools. Each of the following is just one method for producing an effect. The examples are provided to help you concentrate on developing the essential skills of the graining and marbling processes.

Wood effects: oak

Oak is a hard wood, whose natural colour varies from a rich honey colour to a yellowish brown. Its wide range of beautiful grain markings are dependent on the method of conversion. It has been widely used over the years for **maritime** and building construction purposes, as well as furniture.

This technique involves using dragging and combing techniques and a check/tick roller. Oak graining is generally carried out using oil-based scumbles.

KEY TERM

Maritime: related to the sea.

▲ Figure 6.95 Oak – an example of straight grain

Start with a surface ground out in 08 C 35 solvent-borne eggshell, or a tint of it, depending on whether you wish to produce a medium or light oak effect.

Step-by-step oak effect: method 1

STEP 1 Brush out thinned oak scumble glaze to evenly distribute the colour.

STEP 2 Using a dragging brush, produce firm, sharp grain markings.

STEP 3 Apply pore marks using a check/tick roller with mottler.

Step-by-step oak effect: method 2

STEP 1 Brush out thinned oak scumble glaze to evenly distribute the colour.

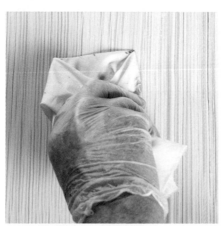

STEP 2 Using a metal comb folded into a piece of lint-free rag, produce firm, sharp grain markings.

STEP 3 With a finer metal comb, use short strokes at a slight angle to the grain marks to break up the vertical lines. (Wipe the tip of the comb regularly to remove excess scumble glaze.)

Wood effects: mahogany

Mahogany is a tropical hardwood of reddish-brown colour, with a fine, silky appearance. Mahogany graining is generally carried out using water graining colour, building up the various characteristics using different processes to create the final depth of rich colour with subtle highlights. However, the process can also be successfully completed using oil-based scumbles.

Start with a surface ground out in 04 D 44 (or 06 C 33) acrylic eggshell.

▲ Figure 6.96 Mahogany – an example of straight grain

Step-by-step mahogany effect

STEP 1 Prepare water graining colour using vinegar/stale beer, water and Vandyke brown.

STEP 2 Degrease the ground coat with Fuller's earth (or whiting).

STEP 3 Evenly brush out the water colour.

STEP 4 Flog the water colour, working away from your body, or upwards if on a door.

STEP 5 Leave to dry (as shown above). Then apply one coat of solvent-borne varnish to bind the water colour to the surface. Allow to dry overnight.

STEP 6 Brush out thinned mahogany scumble over the flogged ground and apply streaks of burnt umber.

The overall finish should be subtle, but effective enough to show some depth when varnished.

STEP 7 Use a fitch brush to spread the colour to produce broad vertical bands.

STEP 8 Using a hog hair softener, lightly soften in a horizontal direction to blur the edges of the bands.

Marble effects

▲ Figure 6.97 A block of quarried Carrara marble

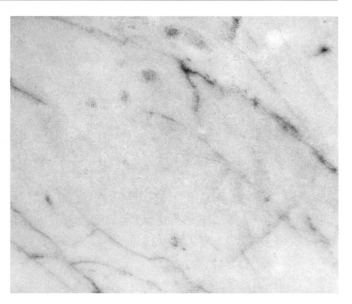

▲ Figure 6.98 Ground and polished marble

Carrara (using softening techniques)

Carrara is a type of white or blue-grey marble, used in sculpture and building decoration. It is quarried near Carrara, in Tuscany, Italy.

Start with a surface that has been ground out in 00 E 55 acrylic eggshell.

Step-by-step process for Carrara marble technique

STEP 1 Prepare a thin, pale grey acrylic scumble using acrylic glaze and white and black colourants.

STEP 2 Apply the acrylic scumble and oil or rub in the surface.

STEP 3 Slightly darken the scumble by adding a small amount of black and apply streaks in a general diagonal direction.

STEP 4 Slightly darken the scumble again and apply darker streaks below the pale ones.

STEP 5 Using a badger softener, follow the general run of the streaks in one direction only, to produce one soft edge and one harder edge, as seen in natural marble.

STEP 6 Further variation can be achieved by applying a plastic bag or clingfilm to the surface. Wipe across the plastic with your hand in a diagonal direction, then carefully remove the plastic from the surface.

STEP 7 Soften to produce the desired effect.

STEP 8 An example of how a finished effect might look.

If the work loses shading definition, re-apply streaks of appropriate greys and work to soften them in. However, as you are working in water-based materials, remember that you will have a limited working time as the material will dry quickly.

The finished work should have the subtlety and depth required of many of the marble types you may imitate.

White marbles such as Carrara may have a yellow shadowy appearance in the background which can be imitated using an overglaze with raw sienna, or a blue cast for which Prussian blue may be used in an overglaze.

ACTIVITY

Produce an example of Carrara marble on card/paper, by carrying out the steps shown. Finish your work with one coat of acrylic varnish to protect it and to prevent the colour from yellowing.

Fantasy marble

The examples show marbling techniques and processes used to produce a realistic-looking, but imaginary, product.

The example shown involves ragging on or sponging as well as oiling in, veining and cissing techniques. Working on a black ground coat with tints and shades of green will produce a black and green 'marble', along the lines of Vert de mer.

Although the following illustrations use an oil-based process it is equally possible to carry out similar processes using a water-based scumble and technique.

▲ Figure 6.99 Vert de mer marble

Step-by-step process for Vert de mer marble technique

STEP 1 Prepare an oil-based glaze using *either* a 2:1 mix of oil-based glaze and white spirit *or* a 2:1 mix of white spirit, linseed oil and driers.

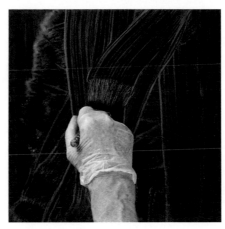

STEP 2 Oil in the surface.

STEP 3 On a palette, prepare a deep green scumble by mixing a small quantity of oil-based glaze with *either* chrome yellow and Prussian blue *or* raw sienna and Prussian blue.

STEP 4 Rag on, or sponge on, the green scumble to the oiled-in surface.

STEP 5 Add white to the green scumble, and rag on or sponge on between the spaces of the previous colour.

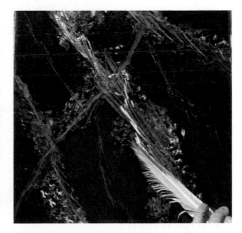

STEP 6 *Either* add more white to the mix of chrome yellow and Prussian blue *or* add Brunswick green and a little white to the raw sienna and Prussian blue. Using the new mix of green, apply a broad vein in a diagonal direction by feather, followed by further veining using various tones of green and pure white.

STEP 7 Soften as necessary.

STEP 8 Ciss the surface by applying tiny splashes of white spirit.

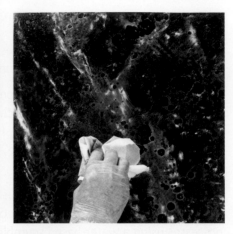

STEP 9 Dab the surface gently to remove excess solvent.

STEP 10 Soften as necessary.

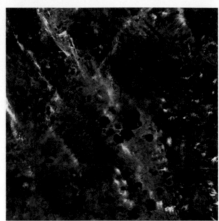

STEP 11 An example of how the finished effect might look.

To enhance the finish and give it greater depth, an overglaze of dark yellowish green (mixed from black, raw sienna, Prussian blue and oil-based glaze) may be applied when the previous processes have dried.

Application processes for structural components

Panelled doors

Follow the same sequence of working as when painting the door, ensuring that all fittings have been removed so that the grain effect is continuous, and therefore more realistic.

1   On the first top panel, rub in the graining colour to the moulding, then the panel itself, brushing the colour out as if applying paint. Remove any excess colour from the corners/moulding detail by lightly stippling with a clean brush to achieve an even distribution of colour. Using the tips of the brush, lay off in the direction of the constructed component parts.

2   Produce the grain detail required by flogging, brush graining (dragging) or combing.

3   Continue to apply the graining colour to the remaining panels one at a time, producing the grain effect immediately and before moving on. Keep adjacent sections (components) as clean as possible by wiping off excess colour.

4 Grain the muntins (Figure 6.78 on page 272 shows the parts of a door), then the rails, finishing each section (component) before moving on to the next. Make sure the joints (butt or mitred) are left sharp and clean.

5 Finally, grain the door edge(s) and stiles.

Any graining colour that may have been left on adjacent sections should be **worked up** if possible, when applying additional graining colour, or wiped off using a lint-free rag moistened with white spirit.

> ### KEY TERM
>
> **Worked up:** re-activated or made workable by rubbing with a brush containing a small quantity of the material, or a suitable thinner.

Particular attention must be given to producing clean, sharp joints which are accurately positioned. This may be achieved by carefully applying masking tape to each end of the muntins before graining them, then removing it as soon as the effect is produced. This method may also be used for the top, middle/lock and bottom rails.

However, when graining the stiles, the previously grained rails need to be protected at the joint so as not to damage the finish; this is usually achieved by holding a 'shield' (such as an oblong piece of plastic/ metal/abrasive paper) over the rail, which will both protect the existing graining and produce a clean, straight joint.

▲ Figure 6.100 Using a shield while graining

Windows

As for panelled doors, work in a methodical manner and in the same sequence that paint would be applied, with the application of graining colour to sections being immediately worked to produce the required effect.

Dado rails and narrow linear runs (architraves and skirtings)

Adjacent surfaces should be protected before work starts, to prevent graining colour marking the wall and to avoid picking up any dirt/dust from the floor, which could contaminate the work.

Due to the length of some of these components, loss of wet edge could be a problem if you are working with water-borne materials. Should this occur and you are unable to adjust the consistency of the material to extend the working time, apply it to smaller sections, joining/blending the new section seamlessly into the existing work.

▲ Figure 6.101 Example of a dado rail

Safety considerations

It is extremely important to be aware of the environmental and health and safety regulations when mixing and using pastes and adhesives. In particular, you must always comply with COSHH Regulations.

Manual handling

Always follow the correct manual handling procedures when carrying out decorating tasks.

Although in most cases decorating involves fairly light work, on occasion materials can be supplied in particularly large and heavy boxes and care should be taken to avoid dropping and perhaps spilling the contents.

If you cannot use a machine, it is important that you maintain the correct posture when lifting any load. The correct technique to do this is known as kinetic lifting. Always lift with your back straight, elbows in, knees bent and your feet slightly apart.

Personal protective equipment

Although most activities involving decorative effects will take place in a closed-room environment, compliance with PPE and site requirements is necessary in order to gain access across a site or within a refurbished property. This means wearing a safety helmet, hi-vis jacket, safety boots, gloves and goggles on some building sites as standard practice. It is usually permissible to remove these while working inside, but always check with your supervisor or site agent first.

Remember to wear the correct PPE and follow safe working practices even when carrying out preparation processes. A face mask may be required if carrying out dry abrading on areas that have been filled with powder-based fillers as there is a risk of inhaling excessive amounts of dust. If you have sensitive skin, wear gloves when decorating. In many cases gloves and barrier cream may be advised to protect the hands from the effects of the solvents. See Chapter 1, page 31 for information on using gloves and barrier cream.

Ventilation

Ensure there is good ventilation in the room as this will allow the paint to dry out naturally and reduce any potential fumes that may be given off.

Good ventilation will reduce the potential build-up of volatile organic compounds (VOCs) that may be contained in the paint (see page 259). Using water-based products will reduce the amount of VOC content released into the atmosphere.

Ingesting chemicals

Ensure that you wash your hands, especially before eating food, to avoid absorbing or ingesting any product, as this could lead to stomach irritation or upset.

Do not smoke or eat when working.

Paints and solvents can be a safety hazard so keep them away from animals and children so that they do not ingest them.

Overalls with a bib are extremely useful as they provide protection and enable you to carry a small number of tools in the pockets.

Safety data sheets

Manufacturers have a legal duty under health and safety legislation to provide information about their products, and it is important to obtain a safety data sheet for each product used. Always ensure that you follow the manufacturer's instructions and carry out a risk assessment using the information provided to determine what the hazards are and how they can be reduced. Advice on first aid measures, such as how to deal with someone ingesting the product, protective equipment required and methods of disposal of waste products, will be covered by the manufacturer's data sheet.

▲ Figure 6.102 First aid kit

material safety data sheet

POLYVINE

Page: 1 Compilation date: 20/02/2019 Revision No: 1
Section 1: Identification of the substance/mixture and of the company/undertaking
1:1 Product identifier
Product name:

ORIGINAL ACRYLIC SCUMBLE
TROPICAL ACRYLIC SCUMBLE

1.2. Relevant identified uses of the substance or mixture and uses advised against
1.3. Details of the supplier of the safety data sheet

Company name:
POLYVINE LTD
VINE HOUSE,
CHEDDAR BUSINESS PARK,
WEDMORE ROAD,
CHEDDAR BS27 3EB
ENGLAND
Tel: 0044 01934 740305
Fax: 0044 01934 744904
Email: laboratory@polyvine.co.uk

1.4. Emergency telephone number

Section 2: Hazards identification
2.1. Classification of the substance or mixture
Classification under CHIP: This product has no classification under CHIP.
Classification under CLP: This product has no classification under CLP.
2.2. Label elements
Label elements: Precautionary statements: P102: Keep out of reach of children. P301+312: IF SWALLOWED: Call
a POISON CENTER/doctor/ if you feel unwell. P262: Do not get in eyes, on skin, or on clothing. P280: Wear
protective gloves and eye protection. P302+350: IF ON SKIN: Gently wash with plenty of soap and water.
P305+351+338: IF IN EYES: Rinse cautiously with water for several minutes. Remove contact lenses, if present
and easy to do. Continue rinsing.
2.3. Other hazards
PBT: This substance is not identified as a PBT substance.

Section 3: Composition/information on ingredients
3.2. Mixtures
There are no ingredients present which, within the current knowledge of the supplier and in the concentrations
applicable, are classified as hazardous to health or the environment, are PBTs, vPvBs or Substances of equivalent
concern, or have been assigned a workplace exposure limit and hence require reporting in this section

▲ Figure 6.103 Material safety data sheet for scumble

ACTIVITY

Source a safety data sheet for an oil scumble glaze,
either by finding one at your college or workplace
or by searching online, and use it to help you
answer the following questions:

1 What safety precautions should you take when
mixing the glaze?

2 Do oil glazes have a high VOC content?

Work at Height Regulations 2005 (as amended)

If any of the work is to be carried out at height, you
must follow the requirements of the Work at Height
Regulations 2005 (as amended). Make sure you assess
all risks and always work safely. Do not stretch higher
than you can reach: use a step ladder or hop-up and
check it is set up correctly before use.

COSHH 2002

Make sure you adhere to the Control of Substances
Hazardous to Health (COSHH) Regulations 2002,
particularly in relation to the handling of materials such
as paints and solvents.

The materials used by painters and decorators need
to be treated carefully to ensure everybody's safety.
In order to keep risks to a minimum, it is important
to know what the risks are. Safety data sheets are
available for the substances you will be using.

The data sheets will provide the following information:
- what the product is
- what it is used for
- identification of hazards
- information on the composition of ingredients
- first aid measures
- firefighting measures
- accidental release measures
- instructions for handling
- guidance for storage.

Disposal of waste

Correct disposal of waste after applying coatings is
very important. Care must be taken when disposing of
hazardous waste and it should never be poured down
the sink. Try not to buy more paint than you need for
a job and consider applying another coat if you have
some left over.

Empty emulsion tins must be washed out and can then
be disposed of in household waste.

It is now possible to buy paint solidifiers, which are
small beads that absorb the paint, turning it into a solid
mass that can be disposed of in the household waste.

Check with your local authority for how to dispose of
oil-based paints and varnishes.

Rags and cloths that have been used to apply chemical
solvents should be fully opened out, allowed to dry
and then disposed of carefully, as they can be a fire
risk if left bound up in a wet state as spontaneous
combustion can occur.

Test your knowledge

1 Which of these is a primary colour?

   a Blue

   b Slate

   c Purple

   d Orange

2 Which of these is a secondary colour?

   a Red

   b Green

   c Blue

   d Olive

3 Which word **best** describes the term 'value' in terms of colour?

   a Lightness

   b Intensity

   c Cool

   d Tinted

4 Which word **best** describes the term 'chroma'?

   a Warm

   b Intensity

   c Cool

   d Neutral

5 Which is a cool colour?

   a Red

   b Blue

   c Yellow

   d Orange

6 Which of these sentences describes the metameric effect under various lighting conditions?

   a Colours appear brighter

   b Colours appear duller

   c Colours appear different

   d Colours appear warmer

7 Which colour combination could be used in an analogous colour scheme?

   a

   b

   c

   d

8 Which colour combination could be used in a complementary colour scheme?

   a

   b

   c

   d

9   What is the most suitable preparation process to use before applying a solvent-borne eggshell ground coat?

   a   Dry abrade using aluminium oxide

   b   Dry abrade using glass paper

   c   Wet abrade using emery cloth

   d   Wet abrade using silicon carbide

10   Why might a mohair roller be used to apply the ground coat for a decorative finish?

   a   To produce a fine textured effect

   b   To reduce the appearance of brush marks

   c   To reduce the number of coats required

   d   To apply the paint quicker

11   A scumble used to produce a broken colour effect is:

   a   transparent, to protect the ground coat colour

   b   translucent, to show the ground coat colour

   c   opaque, to hide the ground coat colour

   d   transparent, to retain the ground coat colour.

12   Why is it a good practice for one operative to apply the scumble and remove brush marks, and a second operative to produce the broken colour effect?

   a   To enable a high work rate

   b   To speed up the drying process

   c   To produce a uniform effect over broad areas

   d   To produce more skilled operatives

13   Which one of the following BS 4800 colours would be suitable to use as a ground coat for oak?

   a   08 C 35

   b   00 E 55

   c   00 E 53

   d   04 D 44

14   When producing a water graining medium, which of the following may stale beer or malt vinegar be used as?

   a   Thinner

   b   Pigment

   c   Drier

   d   Binder

15   Oiling in is a process used for producing which specialist decorative effects?

   a   Graining

   b   Marbling

   c   Sponge stippling

   d   Rag rolling

Practice assignment

Carry out the requirements of the synoptic assignment brief below in the practical workshop.

This section provides a short scenario-based activity that may provide preparation for the synoptic test.

Brief

A client has requested the redecoration of a bedroom. It contains a mixture of features, including a dado rail and panel door. The bedroom has been previously painted and is in good decorative order.

Figure 6.104 gives a basic layout of the room's wall.

The client would like the following:

- Upper wall area: Apply a lining paper and decorate with two coats of a colour of vinyl matt emulsion.
- Dado area: Apply two coats of vinyl silk emulsion, set out border and decorate the centre with two-colour sponge stipple.

The colour scheme to both the upper wall and dado areas should be monochromatic.

The wall and ceiling area are of a standard height, so can be easily accessed using stepladders or hop-ups.

The floor area is carpeted and requires adequate protection.

The client is looking for a professional job, to a high standard of finish, and with no defects.

Health and safety is important and all work should be planned. You should produce a tool and material list together with a method statement detailing what must be done before the project begins in order to ensure that all work is carried out as efficiently and safely as possible.

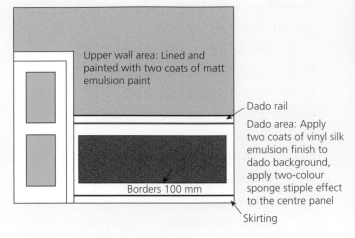

Upper wall area: Lined and painted with two coats of matt emulsion paint

Dado rail

Dado area: Apply two coats of vinyl silk emulsion finish to dado background, apply two-colour sponge stipple effect to the centre panel

Borders 100 mm

Skirting

▲ Figure 6.104 Basic layout of wall

Task 1

Plan to carry out the works. You should produce the following:

- resource list
- method statement.

Task 2

Prepare the upper wall area, hang lining paper vertically and apply two coats of matt emulsion.

Task 3

Prepare the dado area. Apply two coats of vinyl silk emulsion finish to dado background, set out a 100 mm border and apply two-colour sponge stipple effect to the centre panel.

Task 4

Write a reflective evaluation that considers all the tasks you have carried out within the assignment. It is recommended that your evaluation is a minimum of 200 words, or half a page of typed script. Consider what has gone well and what you may change if you were to complete a job like this again in the future.

Glossary

Abrade to remove part of the surface of something by rubbing down with an abrasive material such as silicon carbide paper (commonly known as 'wet and dry').

Abrasion an open wound caused by skin rubbing against a rough surface.

Abrasive the name given to sandpaper or abrasive paper which comes in a variety of grades; the higher the number the finer the grain.

Abrasive papers used when rubbing down to make another surface become smooth. Such materials are very important to a decorator during preparation as inadequate rubbing down of a surface will result in a poor paint finish.

Absorb to take something in, such as through the pores in the skin

Absorbency how much fluid will be absorbed (incorporated) into a material. Highly absorbent surfaces such as softwoods suck up primers and paints, impairing the finish.

Absorbent an absorbent surface soaks up liquids, for example bare plaster, bare timber. The more liquid a material can soak up, the more absorbent it is.

Accent/accentuate in colour terminology, using a small amount of contrast colour will enhance the other colour(s) and add excitement to a scheme.

Access the means to enter or gain entry to a place.

Access equipment equipment used in work at height, such as steps, ladders, podiums, towers, etc.

Adhere to stick to a substance or surface.

Adhesion the action or process of adhering (sticking) to a surface or object.

Adhesive in decorating and particularly paperhanging, an adhesive is a material sometimes referred to as paste that can stick paper to ceiling and wall surfaces.

Aesthetic the beauty of something and how that is appreciated by the person looking at it.

Aesthetically pleasing pleasant to look at.

Alkaline a substance that has a pH greater than 7. Alkalis form a caustic (corrosive) solution when mixed with water. Examples include lime and caustic soda.

Architrave the moulded frame around doors or windows, sometimes referred to as a door frame.

Artex a surface coating used for interior decorating, most often for ceilings, that allows a pattern or texture to be added on application. The name is a trademark of Artex Ltd.

Batch number an identification number used to denote when a batch of wallpaper was produced. It will contain a code indicating the print details. Codes should be the same on all rolls to avoid colour differences.

Bits, nibs or flaking defects caused by dust and poor adhesion of paints.

Building contract an agreement between the client and contractor, which forms a legal agreement of what is included in the work, such as how much it will cost, how payments will be made and the start and completion dates.

Building information modelling (BIM) a digital process for creating and managing information about a construction project throughout its lifecycle, from its earliest conception to completion and potentially its eventual demolition.

Butt joint edges of lengths of paper that touch without a gap or overlap.

Capillarity the rate at which liquid is drawn into a material through pores or small tubes.

Capillary action the ability of water to rise up microscopic tubes in building materials (e.g. most brick types, some stones, concrete blocks and plaster). When these materials are in contact with moisture, the water adheres to the pores of the material's capillaries or microscopic tubes and rises up to cause damp.

Carcinogenic something that can cause cancer.

Castors the swivelling wheels fixed to a scaffold frame.

Centring setting out a wall to create a balanced or even effect for the pattern. Working out from the centre should allow the pattern to be even in appearance.

Chinagraph/grease pencil a writing implement made of hardened coloured wax that is useful for marking on hard, glossy non-porous surfaces.

Chroma the degree of intensity, saturation, purity and brilliance of a colour.

Cissing when a coat of paint or varnish refuses to form a continuous film and leaves the surface partially exposed. The main cause of this is when paint is applied over a greasy surface.

Coalesce when particles merge to form a film, particularly in water-borne coatings, they are said to coalesce. The drying process is known as coalescence.

Coatings paints, varnishes, stains, etc., that are applied to surfaces.

Commercial the term used to describe factories, hospitals, office blocks, etc.

Competent person someone with sufficient training, experience, knowledge and other qualities that allow them to carry out an activity. The level of competence required will depend on the complexity of the situation and in some cases specific training and certification will be required before that person is judged competent.

Concertina fold wallpaper folded like the folds in a concertina musical instrument.

Concisely providing information in a few words, so that communication is brief, but with enough detail to cover all the points.

Coniferous or evergreen trees cone-bearing evergreen trees which keep their leaves in winter.

Consistency refers to what the coating is like in terms of its viscosity (thickness), which can be altered by adding thinners or solvents. In terms of paste, it refers to how thick or thin the material is in use. It is important to have the correct thickness to ensure papers do not become over-soaked.

Contaminants any airborne chemical compounds that affect surfaces, such as adhesives, industrial fallout, rail dust, acid rain, bird droppings, road tar, grime, tree sap, bugs.

Contaminate to pollute or infect.

Contrast/contrasting in colour terminology this usually relates to colours that are opposite on the colour wheel, which go well together.

Corrosion the wasting away of metal when it is exposed to water, oxygen, acid, alkali or salts.

Cotton twill a type of cotton textile that is woven with the characteristic twill pattern of diagonal parallel ribs.

Creep where masking tape has not been securely fixed to a surface and some scumble seeps beneath it – this will result in there not being a sharp edge to the broken colour effect.

Cross lining the process of applying lining paper horizontally as a foundation paper to a vertically hung finishing paper such as patterned paper. This is carried out in some cases where the finishing paper is delicate and also to provide an even, non-porous base, particularly over areas of excess filling or high porosity.

Cure to harden.

Cutting in the process of producing a sharp, neat paint line between two structural components in a room, such as a wall and ceiling or architrave and wall.

Dado an area of wall immediately above the skirting board in a room, and separated from the wall filling by a timber, plaster or plastic strip secured to the wall.

Dado rail the wooden decorative rail that separates the dado from the rest of the wall.

Decant to transfer a liquid by pouring from one container into another.

Decontaminated to clean a substance from a surface.

Defect-free without flaws, holes or cracks and bits left on the surface.

Delamination the separation of the top and bottom layers of paper when over-soaked.

Denatured timber wood that has been exposed to UV light and become grey and friable.

De-nibbed the process of lightly sanding the surface to remove bits and nibs of dust.

Deteriorate to become worse.

Discernible clearly seen.

Domestic the term used to describe people's houses and homes.

Domestic work work carried out in someone's home or property.

Duplex papers made of two layers pressed together for greater strength and to bond the layers together before being sent for printing. The layers are usually embossed to form a pattern in relief.

Dust sheet a sheet used to cover anything that needs protecting from paint or damage.

Dutyholder a competent person that has a duty to provide and maintain a work environment without risks to health and safety. Responsibilities include providing and maintaining safe plant, equipment and structures, safe systems of work and the safe use, handling and storage of plant, structures and substances.

Eco-friendly products that consume less energy and are therefore less harmful to the environment.

Eggshell a type of semi-gloss paint coating.

Egress the means to leave or exit a place.

Extractor a machine that removes dust and fumes from a work area.

Exudation oozing; the release of a liquid.

Figurework/figure graining the main design or pattern of a wood grain, which differs between tree species and according to the method of conversion.

Fire retardant a substance that is used to slow or stop the spread of fire or reduce its intensity.

First fix the main elements of construction, for example roofing, flooring.

Flash point the temperature at which a material gives off a vapour that will ignite if exposed to flame. Chemicals with a low flash point are labelled as highly flammable.

Flashing a defect that occurs in flat and eggshell finishes; it looks like glossy streaks or patches.

Flush level with the rest of the surface.

Foundation in the context of paperhanging, this means providing a suitable base for further treatment, such as hanging finishing paper over the top or for paint application.

Friable easily crumbled or reduced to powder.

Fugitive colours colours that fade when exposed to light. Some colours that are reasonably stable when used at full strength develop fugitive tendencies when mixed with white to create a lighter shade.

Fungicide a substance that destroys fungi.

Generic members of a group or type rather than the name of a specific brand.

Graining applying and manipulating an appropriately coloured scumble to imitate the appearance of a specific timber.

Ground the first coating applied when producing printed wallpaper – a colour that is applied to provide a background.

Grounding out applying the ground coat for painted decorative work.

Halogen an incandescent lamp that has a small amount of halogen gas combined with a tungsten filament to produce a very bright white light.

Hard dry describes a paint film that is hard enough to be worked on without damaging its finish.

Harmony/harmonious terms often used in the description of colour schemes to express that something is pleasing to look at because the colours look good together.

Hazard anything that can cause harm.

Hierarchy the system in which members of an organisation are ranked according to relative status or authority.

Hue a pure colour such as red or yellow.

Human traffic the term used to describe a number of people moving about in or passing by an area.

Humidity moisture in the air.

Hydrophilic describes a coating or material that attracts water and tends to absorb it.

Hydrophobic describes a coating or material whose surface repels water. Droplets hitting a super-hydrophobic coating can fully rebound.

Hygroscopic tending to absorb moisture from the air.

Improvised made or invented using whatever is available, based on an understanding of what is required.

Industrial the term used to describe factories, bridges, etc.

Inertia-operated anchor device a safety device attached to a safety line that works in the same way as a car seat belt to lock when weight falls on it.

Infrastructure the basic systems and services, such as transport and power supplies, needed for people to live and work effectively. Mains power, gas, water, drainage and communications are laid to a construction site to enable the various connections to be made by the electricians and plumbers, etc.

Ingest to take into the body or mouth by swallowing or absorption.

Ingrain commonly referred to as woodchip. Chips of wood are sandwiched between two layers of paper.

Intaglio a design incised or engraved into a material.

Integrity truthfulness or accurate characteristics.

Intense extreme, very strong or having a high degree of colour.

Intumescence relates to the process of a paint coating swelling up or enlarging. As the paint coating swells, it slows down the spread of flame and provides a surface that does not burn easily. There are two main types of intumescence: soft char, which reduces heat transfer by being a poor conductor of heat, and hard char, which is produced with graphite and is used for external fire-proofing.

Inverted turned upside down.

Ironmongery items such as door handles, locks, latches and hinges that are usually made from iron, steel, aluminium, brass and plastics.

Jeopardised when someone or something is put into a situation where there is a danger of loss, harm or failure.

Key a surface that is naturally porous or has been roughened by abrading, to help paint adhere to it.

Key into to seal the surface to allow further coatings to adhere to it.

Kinetic lifting a method of lifting items where the main force is provided by the operative's own muscular strength. Using the recognised technique will avoid injury.

Laying off using the very tips of a brush to lightly stroke the surface of the paint or scumble to minimise brush marks.

LED (light-emitting diode) an electronic device that emits light when an electrical current is passed through it.

Lifecycle the entire time that something exists. For a building, this includes its design, construction, operation and eventual disposal.

Linear relating to straight, often narrow, lines.

Linseed oil a pale yellowish oil made from the seeds of the flax plant and used as a drying oil or binder. Synthetic alkyd resins are more often used nowadays, as they do not yellow as quickly as linseed oil.

Listed building a building that can only be restored, altered or extended if consent is given under government planning guidance. A listed building is given a grade which indicates its level of special interest or significance.

Making good checking a surface for defects such as holes and cracks and applying fillers to make the surface smooth and without defect.

Manipulate to skilfully handle or move.

Manual handling to transport or support a load (including lifting, putting down, pushing, pulling, carrying or moving) by hand or bodily force.

Maritime related to the sea.

Mastic an acrylic type of caulk applied using a mastic gun (also known as a skeleton or caulking gun).

Medium the liquid ingredient that enables a glaze to be spread over a surface and dry as a film. A medium binds the colourant particles together and provides good adhesion to the substrate.

Megilp a variety of substances such as soft soap or beeswax that used to be added to graining colour to make it hold up or prevent it flowing together after it had been combed or figured.

Method of conversion the way a newly felled tree is sawn up into usable-sized pieces of timber, using two main methods: through-and-through or quarter-sawn.

Micro-porous paint a paint with a breathable film that allows moisture and air to be released but prevents moisture such as rain getting in.

Mildew sometimes referred to as mould, it is a fungus that produces a superficial growth on various kinds of damp surface. Mildew is typically seen as black spots that multiply.

Mill a place where paper is produced in its plain form before sending to the manufacturer for processing, in this case to be made into wallpaper.

Monochromatic all the colours (tones, tints and shades) of a single hue.

Mortar the material used to fix bricks together during the building of structures, etc.

Mottler a brush consisting of a ferrule and hog hair bristle filling. It is usually used to produce the characteristic variations of light and shade seen in natural wood.

Mould mould is made up of airborne spores that can multiply and feed on organic matter in pastes (starch pastes contain organic products such as wheat). Mould typically shows as black spots on the surface of paper.

Niche a shallow recess in a wall to hold a statue, vase, etc.

nm nanometre – a unit of measurement used to measure the size of nanoparticles, i.e. how thick nano coatings are, and other very tiny dimensions, such as atoms. A nanometre is one-billionth of a metre.

Non-compliance not following the requirements of legislation, for example the Health and Safety at Work Act, and potentially working in an unsafe manner.

Notation text or numerical references that indicate the groups or categories of colours.

Obliterate to fully cover up/obscure.

Opacity the power of the pigment to obliterate (hide) the existing surface colour.

Opaque not able to be seen through; not transparent. Opacity is the quality of being opaque.

Operative another term for worker.

Overgraining the process of simulating subtle lighter and darker areas of the grain as seen in natural wood – this is carried out, usually in watercolour, over completed figurework.

Paint system the name given to the coatings used such as primer, undercoat and gloss; this is the common paint system used for new skirting boards and architraves, etc.

PASMA Prefabricated Access Suppliers' & Manufacturers' Association Ltd – the lead trade association for the mobile access tower industry.

Paste-the-wall technique the wall is pasted rather than the back of paper.

PAT (portable appliance test) a safety test that must be carried out on electrical items to ensure they are safe to use.

Patching up filling bigger-than-usual holes and missing plaster from the surface before coatings are applied.

Pattern match the pattern match helps you identify where the pattern at the edge of one piece of wallpaper fits together with another roll. This could be an offset match, such as a half-drop or random match, or a straight match. The type of pattern match is given by the symbol on the packet.

Pattern repeat the distance between a single point on the pattern and the next point where it is repeated on the pattern.

Pie chart a circle divided into sectors with each slice representing a proportion of the whole.

Pinholing a pore-like penetration that is present in paints and coatings due to moisture, air, solvents or other fluids being trapped.

Pole sander a hand tool that allows the user to reach higher areas when abrading surfaces.

Porosity the state of being porous – small spaces or voids in a solid material that enable it to absorb liquids.

Porous a solid material having small spaces or voids that enable it to absorb liquids.

Precast beams T-shaped concrete beams or bars are manufactured in the factory workshop (a reinforcement bar is embedded in the beam for strength) in a shape to allow concrete floor blocks to be placed within the spaces when fixed on site.

Precautions measures taken in advance to prevent something dangerous happening.

Preparatory in the context of paperhanging, this refers to the preparation of a surface for further treatment.

Primary first in a sequence. In the case of primary colours, red, yellow and blue are deemed primary as from these first colours many others can be mixed.

Primer the first coat of paint applied to a surface. The main purpose of a priming coat is to make the surface suitable to receive further coats.

Proud where the filler is applied so it is slightly higher than the surface; once dry, it is rubbed down to create a smooth, flat surface.

Proprietary strainers manufacturers' ready-made strainers provide an open mesh that filters out the bits, leaving clean material that is free from contamination.

Psychological relates to the mind, for example the way in which our minds react to a specific colour.

Pulp colour printed directly on an untreated paper (no ground colour applied). Pulp is a cheap form of simplex wallpaper produced in its simplest form from wood pulp. It is often used to describe thin, inexpensive printed wallpapers. These are single layer without the luxury of protective layers such as vinyl and are therefore easily marked.

PVA (polyvinyl acetate) a resin used in both adhesives and paints to provide a hard, strong film. When used as an adhesive, the film is clear and does not stain.

Rag rolling involves the rolling of lint-free rags or a chamois leather into a rough ball and rolled around in a random fashion over the surface. The coloured glaze is manipulated and takes the form of crushed velvet. Avoid using in straight lines as a banded appearance is likely to be formed.

Raked out removal of any loose or crumbling substrates.

Reasonably practicable sensible, taking into account the trouble, time and money involved.

Rebate a rectangular area removed from the corner of a timber section. Used to locate a pane of glass while a frame rebate locates a window sash.

Refract to deflect light from a straight path.

Regulations laws and rules which have been put in place by government and must be followed.

Reinforcement the action or process of strengthening.

Replicate to make an exact copy of.

Risk assessment an assessment of the hazards and risks associated with an activity, and how to reduce and monitor them. The aim of the risk assessment is to identify work hazards and put in place measures to reduce the risk of that hazard occurring.

Ropiness a surface finish defect similar to brush marks, but where the marks are much heavier and coarser; being more pronounced, they are highly visible and unsightly.

Rust refers specifically to the corrosion of ferrous metals.

Sap a sticky, glue-like material which is sometimes known as the blood of the tree.

Scaff tags tags that are attached to commercial tower scaffolding and used to confirm whether the scaffold has been amended and if it is fit to be used. A white tick on a green background indicates the scaffold has been inspected and is fit for use.

Scale the relation between the real size of something and its size on a drawing. For example, 1:1 is full size, while 1:5 means the drawing is one-fifth of the size of the real object.

Screed a levelled layer of material (e.g. cement) applied to a floor or other surface.

Scumble a glaze (translucent product which will retain a design), to which a colourant has been added.

Seal to apply a coating such as primer or knotting solution to the surface of a material to act as a barrier and provide protection.

Secondary second, or the second stage (after primary).

Shading visually checking the shade or colour of wallpaper rolls. Batch and shade numbers should be the same on every roll.

Simplex a wallpaper made from a single layer of paper. It can be produced in smooth or embossed patterns.

Sinking reduction in the sheen of a paint film. This may occur when a section of making good has not been spot-primed and the film former has been partly absorbed by the porous filler.

Size a thin coat of glue or thinned paste applied to an absorbent surface before hanging wallpaper. Size helps to even out the absorbency so that papers do not stick too soon.

Sparingly using very little.

Spontaneous combustion an internal reaction causes an increase in temperature which results in a fire starting without any outside influence.

Spot-primed to apply appropriate primer to sections of surface area that have been made good, to prevent the next coat from sinking into the filler.

Steeped soaked in liquid.

Strain to pour a coating through a porous or perforated device or material in order to separate out any lumps or bits.

Substrate a name used in industry to describe an underlying surface and surface type, such as timber, metal or plaster, onto which paint is applied.

Surface tension the tendency of the surface of a liquid/solid to resist an external force.

Surfactant a compound that lowers the surface tension between two liquids or between a liquid and a solid, to increase adhesion.

Sustainability ensuring the world's natural resources are not used up today, thereby leaving nothing for future generations. Oil and gas reserves are limited, and alternative sources of energy must be found before they are all gone. Trees cut down for wood must be replaced with new trees so that sources of wood do not run out.

Swatch a collection of paint, wallpaper or fabric samples, usually collected into a book form.

Sweat a defect in which paint or varnish develops tackiness or thickens when it is left standing for long periods.

Tactile relating to the sense of touch.

Tertiary third, or the third stage (after secondary).

Tonal balance this is achieved by manipulating the use of colour. For example, a small amount of bright colour can offset the visual weight of a large area of less bright colour. Similarly, a small area of warm colour can balance a large area of cool colour.

Translucent allows light to pass through but things cannot be seen clearly.

Transparent easily seen through, like clear glass.

Tread section of steps that the feet stand on.

Uniform regular; always the same in shape/form, quality, quantity, etc.

Unique the only one of its kind.

U-values a measure of how quickly heat will travel through the floor leading to loss of heat. Different materials have higher or lower U-values and therefore their selection will depend on what they are being used for.

UV light ultraviolet rays, usually from the sun, which can cause health problems (e.g. with the skin or eyes) and damage to materials.

Value in the Munsell system, colour's relative lightness or darkness.

Veneer a thin decorative covering of fine wood applied to a coarser wood or other material.

Viscosity the ability of a liquid or coating to flow; the more viscous it is, the slower it flows.

Visible spectrum the colours of a rainbow: red, orange, yellow, green, blue, indigo and violet.

Volatile organic compounds (VOCs) emitted as gases from certain solids or liquids. VOCs include a variety of chemicals, some of which may have short- and long-term adverse health effects.

Window reveal the sides and head of the window, usually recessed in from the wall.

Worked up re-activated or made workable by rubbing with a brush containing a small quantity of the material, or a suitable thinner.

Improve your maths answers

Chapter 4 (page 181)

8 litres, £84.40

Chapter 5 (page 218)

5 rolls for ceiling if hung in 3 m direction, 6 rolls if hung in 6.2 m direction, and 10 rolls for 4 walls.

Chapter 5 (page 229)

1 Add together all sides of the room to get the perimeter: 4.8 + 4.8 + 4.2 + 4.2 = 18 m
2 Divide perimeter by width of a roll of paper to calculate number of lengths = 18 m ÷ 0.5 m = 36 lengths required

3 To find out how many lengths can be cut from a roll, allowing an additional 100 mm for trimming, overall height becomes 2.4 m. Divide this figure into the length of a roll: 10 ÷ 2.4 = 4.17 (4 lengths per roll)
4 Divide total number of lengths by number of lengths from a roll: 36 ÷ 4 = 9 rolls

Chapter 6 (page 271)

1 5 litres
2 £66.64 plus VAT of £13.33 = £79.97
3 £254.95

Test your knowledge answers

Chapter 1

1	a	6	b
2	b	7	c
3	c	8	a
4	d	9	c
5	c	10	b

Chapter 2

1	a	7	c
2	c	8	a
3	b	9	b
4	d	10	d
5	c	11	c
6	b	12	a

Chapter 3

1	a	6	d
2	d	7	c
3	d	8	a
4	c	9	c
5	b	10	b

Chapter 4

1	c	6	d
2	d	7	c
3	c	8	a
4	b	9	b
5	d	10	c

Chapter 5

1	d	11	d
2	b	12	b
3	c	13	c
4	d	14	c
5	b	15	a
6	c	16	a
7	b	17	c
8	b	18	a
9	c	19	b
10	a	20	b

Chapter 6

1	a	9	d
2	d	10	b
3	a	11	b
4	b	12	c
5	b	13	a
6	c	14	d
7	a	15	b
8	b		

Index